International Conference on Mathematical Physics and Computational Simulation (CONF-MPCS 2023)

Journal of Physics: Conference Series Volume 2669

Oxford, United Kingdom
12-18 August 2023

ISBN: 978-1-7138-8541-2
ISSN: 1742-6588

Printed from e-media with permission by:

Curran Associates, Inc.
57 Morehouse Lane
Red Hook, NY 12571

Some format issues inherent in the e-media version may also appear in this print version.

Institute of Physics
Dirac House, Temple Back
Bristol BS1 6BE UK

Phone: 44 1 17 929 7481
Fax: 44 1 17 920 0979

techtracking@iop.org

Additional copies of this publication are available from:

Curran Associates, Inc.
57 Morehouse Lane
Red Hook, NY 12571 USA
Phone: 845-758-0400
Fax: 845-758-2633
Email: curran@proceedings.com
Web: www.proceedings.com

TABLE OF CONTENTS

Journal of Physics: Conference Series **2669** (2023) 011001 doi:10.1088/1742-6596/2669/1/011001

Committee Members

CONF-MPCS 2023

General Chair

Roman Bauer, *University of Surrey*

Organizing Chair

Marwan Omar, *Illinois Institute of Technology*

Organizing Committee

Sharidya Rahman, *Monash University*
Abdullahi Arabo, *University of the West of England*
Büşra Oğuzhan, *Çukurova University*
Selda Kapan Ulusoy, *Erciyes University*
Mazhar Javed Awan, *University of Management & Technology*
Arshad Hassan Khan, *FAST NUCES Islamabad*

Technical Program Chair

Stavros Shiaeles, *University of Portsmouth*

Technical Program Committee

Achintya Haldar, *University of Arizona*
Bhupesh Kumar, *University of St Andrews*
Altaf Khan, *University of Mianwali*
Yazeed Ghadi, *Al Ain University*
Jie Zhang, *University of Bath*
Mustafa Istanbullu, *Çukurova University*
Gueltoum Bendiab, *University of Frères Mentouri, Constantine*

Publicity Committee

Festus Adedoyin, *Bournemouth University*
Xiaolong Li, *Beijing University of Posts and Telecommunications*

Journal of Physics: Conference Series **2669** (2023) 011001 doi:10.1088/1742-6596/2669/1/011001

Preface

The International Conference on Mathematical Physics and Computational Simulation (CONF-MPCS 2023) is an annual conference focusing on research areas including mathematics, physics, theoretical analysis, and computational simulation. It aims to establish a broad and interdisciplinary platform for experts, researchers, and students worldwide to present, exchange, and discuss the latest advance and development in mathematics, physics, theoretical analysis, and computational simulation.

This volume contains the papers of the International Conference on Mathematical Physics and Computational Simulation (CONF-MPCS 2023). Each of these papers has gained a comprehensive review by the editorial team and professional reviewers. Each paper has been examined and evaluated for its theme, structure, method, content, language, and format.

Cooperating with prestigious universities, CONF-MPCS 2023 organized six workshops in Chicago, Portsmouth, Bath, Abu Dhabi, Adana, and Constantine. Dr. Stavros Shiaeles chaired the workshop "Insider Threat Detection Advances Using AI", which was held at the University of Portsmouth. Dr. Jie Zhang chaired the workshop "Multi-Agent Reinforcement Learning" at University of Bath. And Dr. Yazeed Ghadi chaired the workshop "Enhancing Human-Machine Interaction through Multimodal Interfaces: A User-Centric Approach" at Al Ain University.

Besides these workshops, CONF-MPCS 2023 also held an online session. Eminent professors from top universities worldwide were invited to deliver keynote speeches in this online session, including Dr. Achintya Haldar from University of Arizona, Dr. Roman Bauer from University of Surrey, Dr. Stavros Shiaeles from University of Portsmouth. They have given keynote speeches on related topics of mathematics, physics, and computational simulation.

On behalf of the committee, we would like to give sincere gratitude to all authors and speakers who have made their contributions to CONF-MPCS 2023, editors and reviewers who have guaranteed the quality of papers with their expertise, and the committee members who have devoted themselves to the success of CONF-MPCS 2023.

Dr. Roman Bauer
General Chair of Conference Committee

Peer Review Statement

All papers published in this volume have been reviewed through processes administered by the Editors. Reviews were conducted by expert referees to the professional and scientific standards expected of a proceedings journal published by IOP Publishing.

- **Type of peer review:** Single Anonymous
- **Conference submission management system:** Morressier
- **Number of submissions received:** 138
- **Number of submissions sent for review:** 138
- **Number of submissions accepted:** 32
- **Acceptance Rate (Submissions Accepted / Submissions Received × 100):** 23.2
- **Average number of reviews per paper:** 1.2
- **Total number of reviewers involved:** 7
- **Contact person for queries:**
  Name: Henry
  Email: henry.g@eliwise.ac
  Affiliation: Eliwise Academy Ltd

C/O Ratio Studies in the atmosphere of hot Jupiters

Zhiling Liu

Department of Physics, Beijing Normal University at Zhuhai, 519087, China

202011059058@mail.bnu.edu.cn

Abstract. Exoplanet atmospheres are a window into the chemical composition, internal motion and evolution of extrasolar planets. In the past two decades, more than 5,700 exoplanets have been discovered. The theoretical research and observation of exoplanet atmospheres have developed rapidly. Among the varieties of exoplanets, hot Jupiters are excellent objects for transit observation due to their characteristics of large radius, high temperature and close-in orbit, which provide many valuable research samples for spectroscopy. A large number of observations and studies have been carried out around it. With the rapid development of exoplanet atmospheres studies, this paper aims to provide a brief introduction on the detection methods of planetary atmospheres and the relevant research progress of the carbon-to-oxygen ratio (C/O) of hot Jupiters, which is an important property that reveals the atmospheric processes and planetary evolution, so as to show the prospect of the future study of exoplanet atmospheres.

Keywords: exoplanets, hot Jupiter, planetary atmosphere, C/O.

1. Introduction

The study of exoplanets (extrasolar planets) is one of the fastest growing frontiers of modern astronomy. Early research began in 1995 with the discovery of a Jupiter-mass planet 51 Peg b from observations of periodic-velocity (RV) Doppler wobble in its star [1]. Later in 1996, additional giant planets with short periods were found [2], opening the era of 'hot Jupiter' exoplanets study. The RV technique can reveal the orbital period, semi-major axis, and lower limits of the planet's mass, but no more about themselves. Soon after, the technique of detecting the transiting light curves of star was introduced by Charbonneau et al. in 2000 [3]. With the relative depth of transit curves and the radii of stars, we are able to measure planetary radii, and therefore calculate their densities by combining the mass derived from RV measurements. Up to today, more than 5700 planets have been found [4]. After identifying and locating exoplanets, researchers are able to study their atmosphere using methods which mainly focus on spectroscopy [5]. The atmospheres of planets beyond solar system provide a window to all of an exoplanet's characteristics, including its vast range of exciting physical processes, from supersonic winds and electric currents to strong infrared radiation and exotic chemistry [6].

"Hot Jupiters" are gas giant planets that have typical orbital periods within 10 days around their parent stars. They are known to be close to the star and have considerable sizes, making them easy to be detected by transits, RV or other methods, this is especially true for ground-based transit surveys. Even though, hot Jupiters are still rare objects. So far, 505 hot Jupiters have been detected, accounting for about 9.4% of all confirmed exoplanets [4].

Hot Jupiters are close to their parent stars and receive intense irradiation. As a result, their temperature can reach 1000-3000K [7], this leads to radiation in the infrared. Thus, observations of hot Jupiters mainly focus on the infrared band. Early researches on exoplanetary atmosphere relied mostly on infrared array cameras and spectrometers from HST [8] and Spitzer Telescope [7]. Apart from the majority of H and He similar to the giant planets in the solar system, hot Jupiters' atmospheres also contain other components composed of C, O and N, such as H_2, H_2O, CO and CH_4 [9]. The existence of various molecules is not the only result derived from the spectrum. Another important indicator of atmospheric composition is the ratio of carbon-abundance to oxygen-abundance, i.e., C/O. The value of C/O ratio marks the breaking of chemical equilibrium processes in planetary atmospheres, and provides critical details about their primary origins and subsequent evolution [10].

2. Observational methods

2.1. Transit spectroscopy

Transit spectroscopy is the earliest and most effective method to examine the exoplanet atmospheres. This method focuses on three types of spectra produced by transiting planets as they orbit around their stars, as is shown in Figure 1.: (a) a transmission spectrum when the planet is at primary transit, which is between its star and the observer (b) an emission spectrum when the planet is at secondary eclipse, which is behind its star (c) a phase curve when the planet is between the primary transit and secondary eclipse [5].

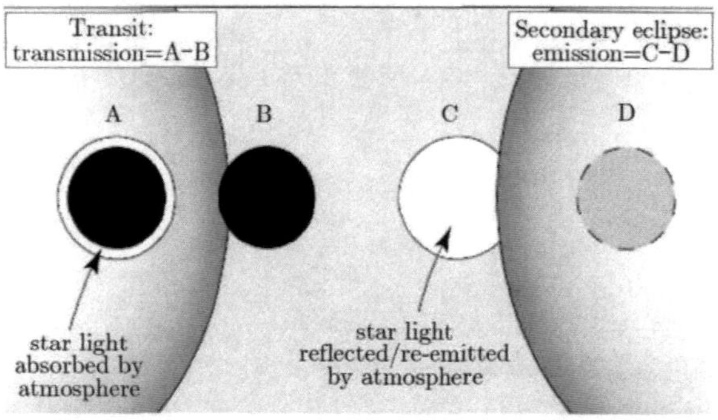

Figure 1. Schematic diagram of primary transit and secondary eclipse [11].

When a planet is at primary transit, the spectrum of the star will show molecular absorption characteristics of different bands because it's blocked by the planet's atmosphere. When a planet is at primary transit, the transit depth τ_1 [11] is

$$\tau_1 = \left(\frac{R_P + A_H}{R_*}\right)^2, \tag{1}$$

where R_P, R^* are the radius of the planet and star, AH is the thickness of the annular absorption band measured at a particular wave band. The transit depth varies at different wavelength, indicating the corresponding chemicals, therefore a set of transmission spectra can be inverted to learn about the planet's atmospheric composition and structure at its terminator region.

When the planet is located behind the star, secondary eclipse occurs. During a secondary eclipse, the transit depth τ_2 [11] is

$$\tau_2 = \left(\frac{T_{eq,p}}{T_{eff,*}}\right)\left(\frac{R_P}{R_*}\right)^2 \tag{2}$$

where $T_{eq,p}$ is the planetary equilibrium effective temperature and $T_{eff,*}$ is the effective temperature of the star. As the planet is much cooler than its star, the planetary spectrum is more pronounced at longer wavelength. By comparing the transit depth before and after the secondary eclipse at different wavelength, we can obtain the planetary reflection spectrum and thermal emission spectrum. The reflectance of the planet and the temperature distribution on its day side can be then derived from inverting these spectrum [11-13].

Planets that have tight orbit around their stars are likely to be tidally locked, resulting in a significant contrast in day and night-time temperatures caused by constant radiative forcing from host stars. By observing the phase curve and analyse its shape and wavelength dependence, we can determine the longitudinal structure, cloud coverage and inhomogeneous chemical composition of the planet [14].

Early studies on transit spectroscopy were mainly based on space telescopes such as Hubble Space telescope (HST) and Spitzer Telescope. In 2002, Charbonneau used the Space Telescope Imaging Spectrograph (STIS) to study the transmission spectrum of HD 209458 and discovered atmospheric Na absorption [13]. Ever since that, numerous studies using transit spectroscopy have been carried out. A large number of molecular spectra have observed [15], and precise measurements on the H_2O abundance of WASP-43b was also delivered [16]. At the same time, more planetary atmospheric structure information such as a map of the planet's atmospheric thermal structure [17] are obtained. Following the success of space telescopes, a number of ground-based telescopes also contributed to transit spectroscopy. With the launch of JWST in 2021, further observations and studies on a great variety of exoplanets including hot Jupiter, are expected to realize [18].

2.2. *High resolution spectroscopy*

High resolution spectroscopy (HRS), or high-resolution Doppler spectroscopy, is another powerful means to directly detect the atmospheres of giant exoplanets at either large and small orbital separations. When collecting the spectrum of a planet, the signals from the star and Earth are almost constant, whereas the planet's spectral lines experience significant Doppler shift. For example, the typical radial velocity of a terrestrial planet is about 30km/s, while the radial velocity of a hot Jupiter is up to 100km/s. Therefore, after high resolution measurements over long time scales, the spectrum of a planet can be separated from the mixed signals by using various detrending methods to remove the signals from stars and Earth atmosphere, and retain the time-varying planetary signals (see Figure 2., the white curves are the planet's atmospheric CO lines, and the dark vertical characteristics are telluric absorption). For close-in planets, the spectrum has a large blue/red shift when orbiting their host stars, which makes detrending easier. For planets with larger orbital radius and lower radial velocity, spectral separation can be achieved by combining HRS with high-contrast imaging. By cross-correlating the remaining spectra after detrending with the template spectra including anticipated significant molecules, it is able to simultaneously monitor of hundreds of molecular spectral lines, and detect various chemical species in exoplanetary atmospheres. By combining the radial velocity and spectrum of the star, the mass of the star and the planet's orbit can be obtained. Coupled with the planet's velocity calculated from the Doppler effect in HRS, the mass of the planet can be accurately determined according to the Newton universal gravitation formula, rather than just the lower limit of mass obtained by the regular RV method.

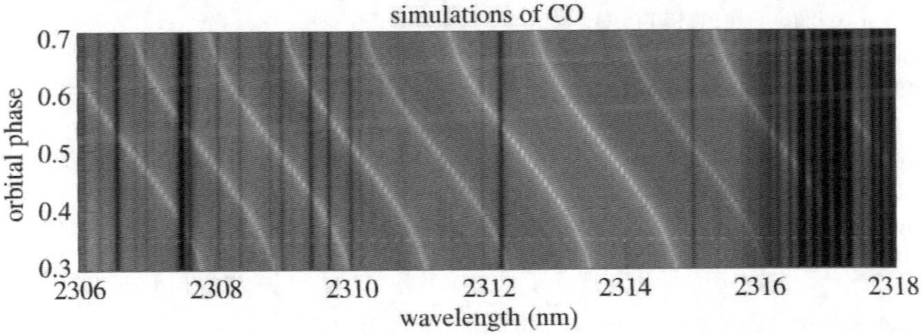

Figure 2. Illustration of CO measurements made by HRS in a hot Jupiter's thermal spectrum[19].

Currently, there are no instruments capable of high-resolution spectral analysis in space. The main tools for HSR are ground-based large telescopes, such as the VLT by ESO. In 2010, Snellen et al. made the first HSR observations of the transiting hot Jupiter system HD 209458b using VLT. When the spectral dispersion R=100000, the carbon monoxide absorption lines in the spectrum decompose into several individual lines around 2300nm and shows significant blue shift within the 5 hours of observation, thus allowing the determination of the mass of stars and planets. At the same time, a small blue shift of carbon monoxide signal relative to the stellar system reveals that a strong wind probably exists in the upper atmosphere of the planet. The proportion of CO in the upper atmosphere of the planet was also estimated by comparing models with different molecular ratios at different temperature [20]. Since then, CRIRES has also been used to study the orbital motion and atmospheric inversion of the planet τ Boötis b, extending the study of atmospheric characteristics to non-transiting planets [21]. Molecules more complex than CO have also been studied, such as H_2O, HCN, etc. [22,23]. Other observatories also conducted studies with HRS using spectrographs such as the HDS on Subaru Telescope, the GIANO on Galileo Telescope, and the NIRSPEC on Keck Telescope. New high-resolution spectrometers, such as ESPRESSO on VLT, and the E-ELT under construction, will promote the efficiency and accuracy of detecting molecules even in terrestrial planets.

2.3. Direct imaging

Direct imaging of gas giants is another method to study exoplanet atmospheres. This method takes direct images of the planet, picking up the visible light it reflects or its own thermal infrared emission. Direct imaging puts stringent requirements on the techniques. A Jupiter analogue at 10pc is too difficult to sperate from its star due to their close angular distance. Meanwhile, young($\lesssim$100Myr) giant planets that have large mass ($\gtrsim$1000m$\oplus$), wide orbit ($\gtrsim$5 AU) and high temperatures($\gtrsim$1000K) [5, 24] enables them to produce near-infrared radiation strong enough to separate from their stars, making them suitable objects for direct imaging. By using a coronagraph, we can remove the influence from the star and get a spectrum of the planet. The resolution and SNR can be improved by using ground-based adaptive optical telescopes with large aperture. Moreover, post-processing can minimize residual aberration, interferometry can further improve angular resolution, and the zeroing interference approach can eliminate stellar light. Compared with transiting spectroscopy, direct imaging detects planets with larger orbital semi-major axes. The greater distance between these planets and their host stars makes them less affected by the intense stellar radiation, which is often the case for transit resulting in very different atmospheric temperature distributions.

Despite the intuitive principles, the number of exoplanets found by direct imaging is just over 200, which is modest compared with the more 3700 planets found by transit, due to the demanding subjects and the restrictions of ground-based observations. After obtaining the spectrum, the analysis is similar from transit spectroscopy and high-resolution spectroscopy. In the visible band, the spectrum of the planet contains absorption bands such as H_2O, O_2 and O_3. Information about the exoplanet's temperature, diameter, mass, surface gravity, and atmosphere pressure may be discovered by examining the visible band's brightness and fluctuation, polarization, and spectral line features. In the infrared band, important molecular bands such as CH_4, CO_2, H_2O and O_3 are important biosignatures to identify the characteristics of extrasolar life [11].

Direct imaging studies of exoplanets began in around 2008, when Kalas et al. [25] and Marios et al. [26] used the Keck telescope respectively to obtain optical images of Fomalhaut b and HR 8799 systems. Over the past two decades, SPHERE on VLT, GPI on Gemini and SCExAO on Subaru have also contributed many high-quality spectra. In the future, space programs such as JWST and WFIRST by NASA, as well as 25-40m class telescopes under construction, such as E-ELT, TMT and GMT, will achieve higher sensitivity, higher resolution and wider spectral range for direct imaging and spectral observation [5].

3. The Carbon-to-oxygen ratio in hot Juptiers' atmosphere

Comparing to the gas giants in the solar system, hot Jupiters have temperatures high enough to allow almost all volatile species to be found in the vapor phase, making them excellent laboratories for the abundance of giant planet atmospheres.

In 2011, Madhusudan et al. reported that, the observed concentrations of CO, H_2O and CH_4 in the transit emission spectrum of the transiting hot Jupiter WASP-12b are consistent with the theoretical expectation of $C/O \geq 1$ [27]. They put forward a theory that planet with $C/O > 0.8$ has an interior dominated by carbides, not silicates like Earth. Oberg et al. suggested that the mechanism by which the planetary C/O ratio deviates from stellar value (such as the solar value of 0.54) may be the different oxygen- and carbon-rich snow lines in the protoplanetary disk. The atmospheric C/O ratio changes systematically as a result of this variation in snow lines [28], which reflects the relationship between C/O and giant planet formation. In 2012, Madhusudan showed that in an atmosphere dominated by hot hydrogen, the carbon-to-oxygen ratio severely affects the relative concentrations of a few spectrally prominent species that cause thermal inversion. Therefore, he proposed to use C/O ratio as a new parameter to characterize exoplanet atmosphere, forming a two-dimensional classification system of radiation (or temperature) - C/O, rather than classifying planets solely according to incident stellar irradiation, making predictions based on solar abundance [29]. This view is widely accepted. But there are still some problems with this theory's prediction, where CO is dominant in the atmosphere of any hot Jupiter. In 2013, Moses et.al studied the spectra of four hot Jupiters and discovered the strong connection between the level of C/O relative to 1 and the different dominant molecules containing carbon or oxygen, which at the same time has little impact on other key species that don't contain carbon or oxygen. This reflects how C/O affects the thermochemical equilibrium of planetary atmospheres [30]. The planetary classification system by Madhusudan has also been further questioned in 2019 by Drummond et al. [31], whose results of one-dimensional atmospheric simulations show that the full abundance of individual elements should also be introduced, rather than only considering the ratio of two elements (e.g., C/O ratio). In 2015, Kreidberg et al. obtained the precise evidence of H_2O absorption in the near-infrared transmission spectrum of transiting planet WASP-12b, which marked the first high-confidence spectral identification of molecules in the atmosphere of WASP-12b, and a new milestone for the confidence of planetary C/O obtained from the spectral inversion [32].

In the recent decade, the study of C/O ratio in exoplanetary atmosphere continues to expand actively. In 2014, Helling et al. proved that the formation process of the planetary clouds would promote the conversion of oxygen-rich atmospheric gas into carbon-rich gas, forming planets with C/O > 1 [33]. Mordasini et al. (2016) discussed the dominant role of envelope enrichment formed by planetesimals impacts on atmospheric composition and C/O ratio of hot Jupiter, which also contributes to the planets with C/O > 1[34]. In general, a planet's C/O ratio can be above 1 only if carbon is produced or inserted into the system in some other way (e.g., planetesimals inflow), or oxygen is reduced to a level below the initial carbon abundance (e.g., used to form clouds) [35]. The model by Molière et al. (2015) pointed out that the photospheric pressure level of self-radiating exoplanets has a strong influence on the molecules that determine opacity, which also corresponds to different level of C/O ratio [36]. They and Cridland et al. (2019) demonstrated that the formation of hot Jupiter usually starts near the water ice line, and its C/O is mainly determined by gas accretion [37]. Based on the study of 10 planetary systems, Brewer et al. (2017) predicted that the C/O of most hot Jupiters would be higher than that of their host stars [38]. In 2019, Helling et al. found that the increase of C/O value was an obvious sign of cloud formation when studying the cloud properties of superhot Jupiters, making it necessary to introduce cloud modeling in the tracer study of planet formation [39]. In 2023, Polman et al. found that H_2S and SO_2 are easy to be observed at high C/O and low C/O respectively. The recently launched JWST is expected to detect these two species, and provide detections at higher accuracy, testing more hypotheses than ever before [40].

4. Conclusion and future landscape

Hot Jupiter is one of the most important objects in the study of exoplanets. Because of its large radius, high temperature and small orbital radius, it is ideal for transit spectroscopy and high resolution spectroscopy. After more than two decades of development, the study of exoplanet atmosphere has made great progress. From the early reliance on the HST and Spitzer space telescopes, to the application of large-aperture ground-based telescopes such as Keck, Gemini and VLT in high-resolution spectroscopy and direct imaging, to the recent launch of the JWST telescope, the tools for exoplanetary atmosphere research are increasingly abundant. The theory and research of hot Jupiters' atmosphere, especially concerning the value of C/O ratio in hot Jupiter, its atmospheric processes and planetary evolution are constantly updated. In this paper, the development and primary research techniques for exoplanet atmospheres are outlined, along with a summary of relevant studies on the C/O ratio in the atmosphere of hot Jupiter. In the future, JWST, WFIRST mission and other 25-40m telescopes such as E-ELT, TMT and GMT, as well as the high-resolution spectrometers such as ESPRESSO on the VLT, are expected to bring more samples in higher sensitivity and better resolution to the exoplanet atmosphere studies, and to uncover the mystery of hot Jupiters' formation.

References

[1] Mayor M, Queloz D. A Jupiter-mass companion to a solar-type star[J]. nature, 1995, 378(6555): 355-359.
[2] Marcy G W, Butler R P. A planetary companion to 70 Virginis[J]. The Astrophysical Journal, 1996, 464(2): L147.
[3] Charbonneau D, Brown T M, Latham D W, et al. Detection of planetary transits across a sun-like star[J]. The Astrophysical Journal, 1999, 529(1): L45.
[4] http://exoplanet.eu/catalog/
[5] Madhusudhan N. Exoplanetary atmospheres: Key insights, challenges, and prospects[J]. AnnualReview of Astronomy and Astrophysics, 2019, 57: 617-663.
[6] Deming L D, Seager S. Illusion and reality in the atmospheres of exoplanets[J]. Journal of Geophysical Research: Planets, 2017, 122(1): 53-75.
[7] Charbonneau D, Allen L E, Megeath S T, et al. Detection of thermal emission from an extrasolarplanet[J]. The Astrophysical Journal, 2005, 626(1): 523.
[8] Swain M R, Vasisht G, Tinetti G. The presence of methane in the atmosphere of an extrasolar planet[J]. Nature, 2008, 452(7185): 329-331.
[9] Madhusudhan N, Seager S. A temperature and abundance retrieval method for exoplanet atmospheres[J]. The Astrophysical Journal, 2009, 707(1): 24.
[10] Madhusudhan N, Harrington J, Stevenson K B, et al. High C/O Ratio and Weak Thermal Inversion in the Very Hot Atmosphere of Exoplanet WASP-12b[J]. arXiv preprint arXiv:1012.1603, 2010.
[11] L.Tian, J.Ji. Research status of exoplanetary atmosphere[J]. Progress in Astronomy, 2015, 33(1): 27-43. (in Chinese)
[12] Kreidberg L. Exoplanet atmosphere measurements from transmission spectroscopy and otherplanet-star combined light observations[J]. arXiv preprint arXiv:1709.05941, 2017.
[13] Charbonneau D, Brown T M, Noyes R W, et al. Detection of an extrasolar planet atmosphere[J].The Astrophysical Journal, 2002, 568(1): 377.
[14] Parmentier V, Crossfield I. Exoplanet phase curves: Observations and theory[J]. arXiv preprintarXiv:1711.07696, 2017.
[15] Vidal-Madjar A, Désert J M, Des Etangs A L, et al. Detection of oxygen and carbon in the hydrodynamically escaping atmosphere of the extrasolar planet HD 209458b[J]. The Astrophysical Journal, 2004, 604(1): L69.
[16] Kreidberg L, Bean J L, Désert J M, et al. A precise water abundance measurement for the hot Jupiter WASP-43b[J]. The Astrophysical Journal Letters, 2014, 793(2): L27.
[17] Stevenson K B, Désert J M, Line M R, et al. Thermal structure of an exoplanet atmosphere from phase-resolved emission spectroscopy[J]. Science, 2014, 346(6211): 838-841.
[18] Greene T P, Line M R, Montero C, et al. Characterizing transiting exoplanet atmospheres with JWST[J]. The Astrophysical Journal, 2016, 817(1): 17.
[19] Snellen I. High-dispersion spectroscopy of extrasolar planets: from CO in hot Jupiters to O2 in exo-Earths[J]. Philosophical Transactions of the Royal Society A: Mathematical, Physical and Engineering Sciences, 2014, 372(2014): 20130075.
[20] Snellen I A G, De Kok R J, De Mooij E J W, et al. The orbital motion, absolute mass and high-altitude winds of exoplanet HD 209458b[J]. Nature, 2010, 465(7301): 1049-1051.
[21] Brogi M, Snellen I A G, De Kok R J, et al. The signature of orbital motion from the dayside of the planet τ Boötis b[J]. Nature, 2012, 486(7404): 502-504.

[22] Birkby J L, De Kok R J, Brogi M, et al. Detection of water absorption in the day side atmosphere of HD 189733 b using ground-based high-resolution spectroscopy at 3.2 μm[J]. Monthly Notices of the Royal Astronomical Society: Letters, 2013, 436(1): L35-L39.

[23] Hawker G A, Madhusudhan N, Cabot S H C, et al. Evidence for multiple molecular species in the hot Jupiter HD 209458b[J]. The Astrophysical Journal Letters, 2018, 863(1): L11.

[24] Fischer D A, Howard A W, Laughlin G P, et al. Exoplanet detection techniques[J]. arXiv preprint arXiv:1505.06869, 2015.

[25] Kalas P, Graham J R, Chiang E, et al. Optical images of an exosolar planet 25 light-years from Earth[J]. science, 2008, 322(5906): 1345-1348.

[26] Marois C, Macintosh B, Barman T, et al. Direct imaging of multiple planets orbiting the star HR 8799[J]. science, 2008, 322(5906): 1348-1352.

[27] Madhusudhan N, Harrington J, Stevenson K B, et al. High C/O Ratio and Weak Thermal Inversion in the Very Hot Atmosphere of Exoplanet WASP-12b[J]. arXiv preprint arXiv:1012.1603, 2010.

[28] planetary atmospheres[J]. The Astrophysical Journal Letters, 2011, 743(1): L16.

[29] Madhusudhan N. C/O ratio as a dimension for characterizing exoplanetary atmospheres[J]. The Astrophysical Journal, 2012, 758(1): 36.

[30] Moses J I, Madhusudhan N, Visscher C, et al. Chemical consequences of the C/O ratio on hot Jupiters: examples from WASP-12b, CoRoT-2b, XO-1b, and HD 189733b[J]. The Astrophysical Journal, 2012, 763(1): 25.

[31] Drummond B, Carter A L, Hébrard E, et al. The carbon-to-oxygen ratio: implications for the spectra of hydrogen-dominated exoplanet atmospheres[J]. Monthly Notices of the Royal Astronomical Society, 2019, 486(1): 1123-1137.

[32] Kreidberg L, Line M R, Bean J L, et al. A detection of water in the transmission spectrum of the hot Jupiter WASP-12b and implications for its atmospheric composition[J]. The Astrophysical Journal, 2015, 814(1): 66.

[33] Helling C, Woitke P, Rimmer P B, et al. Disk evolution, element abundances and cloud properties of young gas giant planets[J]. Life, 2014, 4(2): 142-173.

[34] Mordasini C, van Boekel R, Mollière P, et al. The imprint of exoplanet formation history on observable present-day spectra of hot Jupiters[J]. The Astrophysical Journal, 2016, 832(1): 41.

[35] Helling C, Kawashima Y, Graham V, et al. Mineral cloud and hydrocarbon haze particles in the atmosphere of the hot Jupiter JWST target WASP-43b[J]. Astronomy & Astrophysics, 2020, 641: A178.

[36] Mollière P, van Boekel R, Dullemond C, et al. Model atmospheres of irradiated exoplanets: The influence of stellar parameters, metallicity, and the C/O ratio[J]. The Astrophysical Journal, 2015, 813(1): 47.

[37] Cridland A J, van Dishoeck E F, Alessi M, et al. Connecting planet formation and astrochemistry-A main sequence for C/O in hot exoplanetary atmospheres[J]. Astronomy & Astrophysics, 2019, 632: A63.

[38] Brewer J M, Fischer D A, Madhusudhan N. C/O and O/H Ratios Suggest Some Hot Jupiters Originate Beyond the Snow Line[J]. The Astronomical Journal, 2017, 153(2): 83.

[39] Helling C, Lewis D, Samra D, et al. Cloud property trends in hot and ultra-hot giant gas planets (WASP-43b, WASP-103b, WASP-121b, HAT-P-7b, and WASP-18b)[J]. Astronomy & Astrophysics, 2021, 649: A44.

[40] Polman J, Waters L, Min M, et al. H2S and SO2 detectability in Hot Jupiters: Sulfur species as indicator of metallicity and C/O ratio[J]. arXiv preprint arXiv:2208.00469, 2022.

Limit cycle of Fitzhugh–Nagumo equation

Xiang Ding

South China University of Technology, Mathematics major, Guangzhou, China

202130320366@mail.scut.edu.cn

Abstract. In the 1950s Hodgkin and Huxley created a model to describe the electrochemical transmission of neuronal signals along the cell membrane, it's a very important model. However, in view of its complexity, this paper takes the next step to study its simplified model -- Fitzhugh–Nagumo equation. In these equations x is similar to the voltage and represents the excitability of the system, the variable y represents a combination of other forces that tend to return the system to rest. The parameter I is a stimulus parameter that leads to excitation of the system, I is like an applied current. Many studies have been made on the traveling wave solutions and solitary wave solutions of Fitzhugh–Nagumo equation under different parameters, and many people have also studied its numerical solutions. In this paper, the existence and uniqueness of periodic solutions of Fitzhugh-Nagumo equation with different parameters are studied by using geometric analysis and the results of Van der Pol equation.

Keywords: Fitzhugh-Nagumo equation, the uniqueness of periodic solutions, limit cycles, hope bifurcation.

1. Introduction

One of the most important developments in the study of the firing of nerve cells or neurons was the development of a model for this phenomenon in giant squid in the 1950s by Hodgkin and Huxley [1]. The system described the electrochemical transmission of neuronal signals along the cell membrane and it helped they later receive the Nobel Prize [2]. From Edelstein-Keshet [3] or Murray [4], neurons consist of cell bodies, or cell bodies, that receive electrical stimuli, which are then conducted along the axon, which can be thought of as a cable that connects to other neurons through a series of synapses. Although this process is not an actual electric current, it can still be described in terms of a system similar to a circuit system. Although this system is not as biologically accurate as the original system, it nevertheless does capture the essential behaviour of nerve impulses. This system is obviously of great significance for biological research. However, it is difficult to deal with primarily because of the highly nonlinear nature of the equations. So, a simpler model of the Hodgkin–Huxley model named Fitzhugh–Nagumo system was produced (Fitzhugh–Nagumo equation will be abbreviated to FN equation for convenience in the following). At present, FN equations have been studied a lot, and even the analytical and numerical solutions of fractional FN equations have made some achievements [5].

Although FN equation is a simplified version of the original system, the research of the original system can be promoted by discussing some of its properties. Noting the similarity between FN equation and Van der Pol equation, this paper will apply the research methods of periodic solutions of Van der Pol equation in two literatures to the FN equation. In differential equations, dynamical systems, and an introduction to chaos, the existence and uniqueness of limit cycles are proved by means of geometry. In this process, the algebraic part is relatively less, so the proof is simple and intuitive and easier to understand. However, this method has strong limitations. When the number of parameters is large, there will be a lot of classification discussion. The existence and uniqueness of limit cycles of Van der Pol equation is proved by Journal of Weifang University [6]. In contrast, the lemma and train of thought of this proof use more algebraic methods, and the process of number and form combination is well completed, rather than a very pure geometric process.

Qualitative Theory of Differential Equations [7] introduces three important lemmas: Lehner's graph method, PB ring domain theorem and GS theorem. The paper will not prove lemmas in the subsequent proof process. In addition, some trivial conclusions such as the comparison theorem [8] and Newton's iteration formula [9] will be used without proof.

2. Methodology

Van der Pol equation and Hope bifurcation are introduced in the literature, but the FN equation does not go further, but only provides some ideas. This paper is based on the ideas of this book and extends the idea of Van der Pol equation discussed in the previous section to FN equation, so under what conditions the FN equation has a unique periodic solution can be figured out.

$$\begin{cases} x' = y + x - \dfrac{x^3}{3} + I \\ y' = -x + a - by \end{cases} \tag{1}$$

where a and b are constants satisfying

$$0 < \frac{3}{2}(1-a) < b < 1,$$

and I is a parameter.

First assume that $I = 0$. So, the FN equation becomes

$$\begin{cases} x' = y + x - \dfrac{x^3}{3} \\ y' = -x + a - by \end{cases} \tag{2}$$

Thinking the equilibrium point $x' = y' = 0$, then $y_1 = \frac{x^3}{3} - x$ and $y_2 = -\frac{1}{b}x + \frac{a}{b}$.

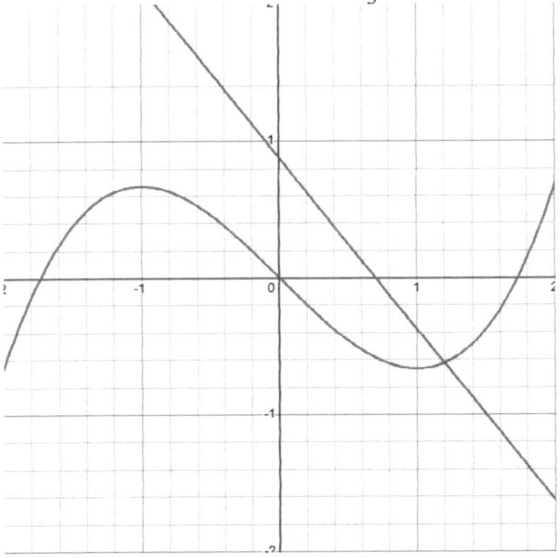

Figure 1. y_1 intersects y_2.

Since $0<b<1$ and $\frac{a}{b}>0$, $y_1 > y_2$ at $x = 0$, y_1 and y_2 must have only one intersection from figure 1, that is, only one equilibrium point (x_0, y_0). Then see the phase portrait when $a = 0.7$, $b = 0.8$ from Wolfram Alpha like figure2:

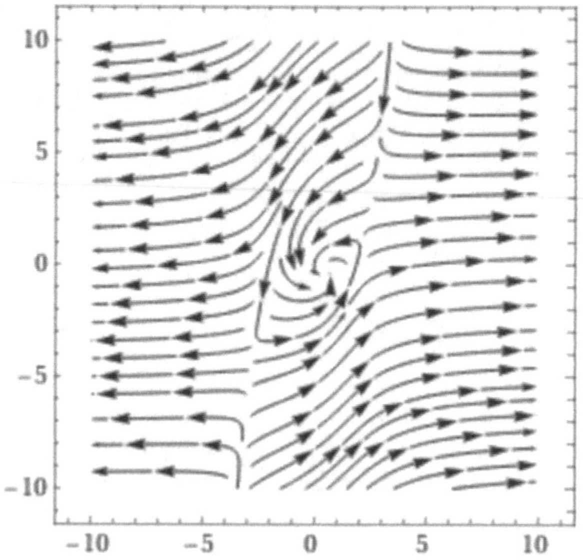

Figure 2. The phase portrait when $a = 0.7$, $b = 0.8$.

This shows that the equilibrium point is always a sink. In fact, this is true for all the other parameters a and b that satisfy the condition. Noting $y = \frac{x^3}{3} - x + \frac{1}{b}x + \frac{a}{b} + I$, then $y' = x^2 + \frac{1}{b} - 1 > 0$, y is continuous, $\lim_{x \to -\infty} y = -\infty$, $\lim_{x \to +\infty} y = +\infty$, so y has one and only one zero on R which means y_1 and y_2 must have only one intersection.

Now suppose that $I \neq 0$, It's obviously going to have the same result: there is still a unique equilibrium point (x_1, y_1). Then $y_3 = \frac{x^3}{3} - x - I$ and $y_4 = \frac{x^3}{3} - x$. It is obvious to imagine from the figure 3 that as I monotonically increases or decreases, y_1 will move up or down and become y_3, so that the x-coordinate value of the intersection of y_3 and y_4 will also monotonically change with I. For example, let $I = -1$, $I = -2$ and draw the portrait when $I = -1$, then the equilibrium point become a source like figure 4.

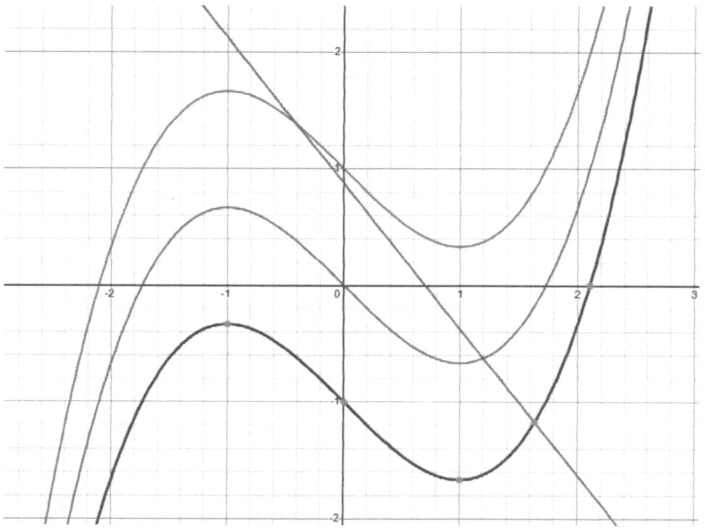

Figure 3. The graph where y_1 intersects y_2 when $I = 0, -1, -2$.

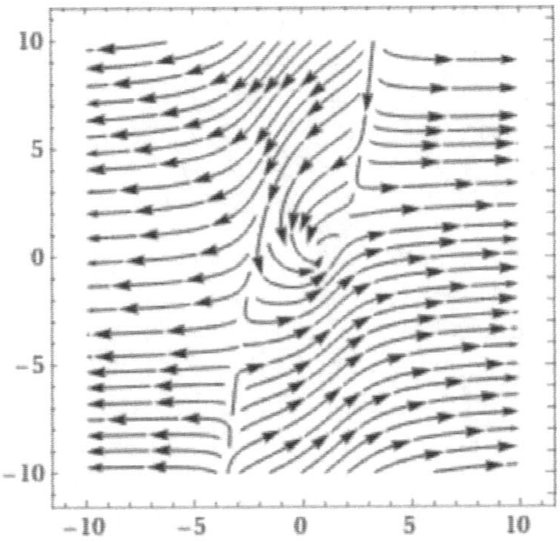

Figure 4. The phase portrait when $a = 0.7, b = 0.8, I = -1$.

When $I \neq 0$, the point (x_0, y_0) is no longer an equilibrium point. Nonetheless, it is possible to consider the solution through this point and to describe the qualitative nature of this solution as I moves away from 0. This idea is in the original literature, but the qualitative property may be too weak, so the author uses Hope bifurcation here to find the limit cycle directly and get a stronger expression directly. Consider the Hopf bifurcation of (1) [10]:

$$\begin{cases} x' = y + \lambda\left(x - \frac{x^3}{3} + I\right) \\ y' = -x + \lambda(a - by) \end{cases} \tag{3}$$

Denote function:

$$\Phi(A) = \int_0^{2\pi} x\left(x - \frac{x^3}{3} + I\right) + y(a - by)\, dt \tag{4}$$

Let $x = Acost, y = Asint$,
Then

$$\Phi(A) = \int_0^{2\pi} Acost\left(Acost - \frac{(Acost)^3}{3} + I\right) + Asint(a - bAsint)\, dt \tag{5}$$

$$= A^2\pi - A^4\frac{\pi}{4} - bA^2\pi \tag{6}$$

Let $\Phi(A) = 0$, then $A = 0$ or $A = \pm 2\sqrt{1-b}$, $\Phi'(A) = 2A\pi - A^3\pi - 2bA\pi$
$\Phi'(-2\sqrt{1-b}) = -4\sqrt{1-b}\,\pi - 8(1-b)^{\frac{3}{2}} + 4b\sqrt{1-b}\,\pi < 0$.
So, the Hopf bifurcation can be derived from the above conclusion.

$$\begin{cases} x' = y + \lambda\left(x - \frac{x^3}{3} + I\right) \\ y' = -x + \lambda(a - by) \end{cases} \tag{7}$$

And it has a limit cycle near

$$\begin{cases} x = 2\sqrt{1-b}\, cost \\ y = 2\sqrt{1-b}\, sint \end{cases} \tag{8}$$

when $\lambda > 0$ it is stable, when $\lambda < 0$ it is unstable.

It can be seen from (8) that the varying of a and I do not affect the bifurcation of the FN equation and only the varying of b should be considered. But from the expression of A, if $b > 1$, then there is no limit cycle on the real phase. If $b \leq 1$, then there is a limit cycle on the real phase, the stability depends on whether $\lambda > 0$ or $\lambda < 0$.

At the same time, since the method of proving the unique periodic solution of Van der Pol equation in the Journal of Weifang University is different from that in the former book, the author will also generalize some ideas of the method in the journal to FN equation. Firstly, three lemmas are introduced.

Lemma1($Li\acute{e}nard$ plotting): Set an equation

$$\begin{cases} x' = y - F(x) \\ y' = -g(x) \end{cases} \tag{9}$$

Take a point $p\ (x, y)$ on the phase plane, make another point $Q\ (0, F(x))$, connect PQ, pass P to make a vertical line of PQ, then this vertical line is the direction field of point P, the direction of which is determined by the original equation.

Lemma2($Poincar\acute{e} - -Bendixson$ ring domain theorem): If there is a ring region G enclosed by two simple closed curves Γ_1 and Γ_2 in the region D, it satisfies the conditions:

i.Γ_1 is in the inner domain of Γ_2

ii. There are no singularities in G

iii. All orbitals that intersect the boundaries Γ_1, Γ_2 of G go from the outside of G to the inside of G as t increases (or decreases)

Then there is a limit cycle in G, which contains the inner boundary Γ_1 of G in its inner domain

Lemma3($G.Sansone$ theorem): Given system of differential equations

$$\begin{cases} x' = y - F(x) \\ y' = -x \end{cases} \tag{10}$$

Where $F(x) = \int_0^x f(s)ds$. If it satisfies:

i.f(x)∈(-∞, +∞); $\exists\delta_1<0<\delta_2$, when $\delta_1<x<\delta_2$, f(x)<0; when $\delta_1>x$ or $x>\delta_2$,f(x)>0,

ii.$\exists\Delta> 0,\ F(\Delta) = F(-\Delta) = 0,$

iii. $\lim_{x\to+\infty} F(x) = +\infty,\ \lim_{x\to-\infty} F(x) = -\infty.$

Then the limit cycle of the equation is unique.

According to the conditions of these three lemmas, it can be seen that the lemma is only true if $b \equiv 0$. In this parameter condition, the existence of limit cycles is first proved. Draw the image of FN equation with different parameters on the phase plane through the first two lemmas, then find out the trend of any point in the phase plane by Newton iteration theorem, and finally prove the existence of the limit cycle of the equation by ring domain theorem. The proof process is almost exactly the same as in the original literature. Parameters a and I only cause translation of the image on the phase plane, so the specific process will not be elaborated here. In addition, the isocline method can be also used to find the direction of the curve field direction, and also use the ring domain theorem to prove the existence of limit cycles. It is proved that the existence of limit cycle of another way, the causes and just similar, concrete proof nor give shows. Finally, lemma three proves that the limit cycle is unique.

3. Discussion

For the first proof method, its disadvantage is very obvious. Although its geometric significance is very strong and it can be well deduced to more advanced equations, this method has very high requirements on parameters. For example, in this paper, if the range of a and b is not limited, if only the pure geometric method is used, even the first step -- whether there is a unique equilibrium point or not -- is very complicated, let alone whether there is a limit cycle in each parameter of the equation, although perhaps the author is not very good at it. But it also shows that this method in the process of promotion may be difficult to apply directly. And when generalized to more advanced equations, the desirable range of parameters of the purely geometric method becomes smaller and smaller, and is of little significance to practical problems. Therefore, in the subsequent discussion of $b > 1$ and the synchronous changes of I and a, the author uses the Hope bifurcation method to directly find out the expression of its limit cycle with algebraic equations, which is more general and more suitable for generalization to some extent.

In the second proof, the limitations are also obvious. Because restricted to three lemmas, the parameter b of FN equation can only be equal to 0, so all the results are based on this condition only. However, when $b \equiv 0$, FN equation is only translated from Van der Pol equation, and this result is only the simplest generalization of Van der Pol equation, which is quite different from the expectation. From the first proof, it is possible for the equation to have a unique limit cycle when b is not zero, which means that this method can only prove the case on a small scale, which is very similar to the case of pure geometry. But the solution is not as complicated as the pure geometric method, if the lemmas can be generalized, so that b is not zero, the result will be more complete naturally, because the idea of proof is almost the same, it's extremely easy to generalize to higher level equations, all have to do is study the lemma, The lemma is improved to fit the parametric case of the equation. In this paper, the author is unable to generalize the lemma due to lack of skill, so this proof method is only presented as a different idea. Another limitation is that this method can only first find the limit cycle under certain parameter conditions, and then prove whether it is unique on this basis. However, if no limit cycle is found, it cannot prove that the limit cycle does not exist. This limitation may not be improved on the previous one and it is not suitable for the study of a completely new equation.

4. Conclusion

In this paper, two methods are used to study when FN equations have limit cycles, and the uniqueness and stability of the limit cycles. In the parameter range given by the original formula, FN equations have unique limit cycles. By using the Hope bifurcation method, when $b > 1$, there is no limit cycle in the original equation, let alone its uniqueness. Method two only deals with what happens when b is equal to 0, which is just a different way of proving it, and knowing that there is a unique limit cycle no matter what.

References

[1] Hodgkin, A. L. & Huxley, A. F. (1952). A quantitative description of membrane current and its application to conduction and excitation in nerve. Journal of Physiology, 117. Doi:10.1113/jphysiol.1952.sp004764

[2] Hirsch, M. W. Smale, S. & Devaney, R. L. (2004). Differential Equations, Dynamical Systems, and an Introduction to Chaos. Academic Press.

[3] Edelstein-Keshet, L. (1988). Mathematical Models in Biology.

[4] Murray, J. D. (1989). Mathematical Biology II.

[5] Tasbozan Orkun. (2021). A popular reaction-diffusion model fractional Fitzhugh-Nagumo equation: analytical and numerical treatment. Applied Mathematics-A Journal of Chinese Universities (2). Doi:10.1007/S11766-021-3810-X.

[6] Xin,Y.Z. (2001). Fande boer fangcheng jixianhuan de cunzaixing[The existence of limit cycles for Van der Pol Equation]. Journal of Weifang University, 1(2), 3. Doi: 10.3969/j.issn.1671-4288.2001.02.003

[7] Zhang,Z.F. (1985). Weifen fangcheng dingxing lilun[Qualitative theory of differential equations]. Kexue chuban she.

[8] You,B.L. (1981). Gaodeng xuexiao shiyong jiaocai Changweifen fangcheng buchong jiaocheng [A supplementary course on ordinary differential equations in a trial textbook for colleges and universities]. Gaodeng jiaoyu chuban she.

[9] Liu,C.K. (1986). Yingyong fanhan fenxi [Applied Functional Analysis]. Guofang gongye chuban she.

[10] Xu,X.P. (2006). Cong hopf Fencha dao jixianhuan [From hopf bifurcation to limit cycle]. Gaoshi like xuekan, 26(4), 5. Doi: 10.3969/j.issn.1007-9831.2006.04.009.

A review of UWB indoor positioning

Junwei Qu

School of Precision Instrumentation and Optoelectronic Engineering, Tianjin University, Tianjin, China, 300072

urraa7@163.com

Abstract. With the development of science and technology, more accurate and stable location information can better serve people's lives. As people spend more time indoors, traditional positioning technologies such as Wi-Fi can no longer meet their needs due to a lack of positioning accuracy and interference resistance. UWB is an emerging positioning technology with high accuracy, robustness, and stability. This paper compares the advantages and disadvantages of UWB technology with those of Bluetooth and Wi-Fi. Common UWB geometric positioning methods such as TOA, TDOA, etc. are introduced. By introducing the latest research progress in UWB indoor positioning, an outlook on the future development of UWB indoor positioning is made. Compared with Bluetooth and other indoor positioning technologies, UWB positioning technology has the advantages of high positioning accuracy and good anti-interference. The UWB positioning system usually uses TOA, TDOA, AOA, and other positioning algorithms. In the future, UWB positioning technology will serve more and more complex positioning environments.

Keywords: UWB, indoor positioning, TOA, AOA, TDOA.

1. Introduction

Along with the development of science and technology, location-based services have entered every aspect of human life. UWB technology has a high potential for solving indoor positioning problems due to its stability and anti-jamming capability.

Ultra-Wide Band (UWB) was approved by the FCC in 2002 for operation in the public frequency band and requires strict compliance with the 3.1GHz-10.6GHz communication band. In the recent 20 years, UWB technology has received widespread attention and experienced rapid development due to its high accuracy, reliability, high resolution, and strong anti-interference characteristics. UWB technology has become a popular method for improving indoor positioning technology [1].

This paper first introduces the working principle of UWB and compares UWB technology with other common indoor positioning techniques, analyzing the advantages and disadvantages of different positioning techniques. The paper then introduces and compares several different geometric positioning techniques and models. Finally, the shortcomings of UWB and future development expectations are analyzed by presenting the latest advances in UWB indoor positioning technology in recent years [1, 2].

2. Indoor positioning technology

With the continuous development of wireless technology, indoor positioning technology and wireless positioning technology continue to combine. On the one hand, indoor positioning technology has undergone rapid development. On the other hand, a large number of wireless positioning technologies are in indoor positioning systems to give full play to their advantages, which are prominent: Bluetooth, Wi-Fi technology, infrared technology, ultra-wide band, etc.. Different technologies have different advantages and disadvantages and are used on different occasions.

Wi-Fi and Bluetooth are currently the most widely used technologies, as they require less equipment. Wi-Fi positioning uses the nearest neighbor method, or cross-location, which is less expensive but only determines the approximate location of the target object [3]. Bluetooth technology, on the other hand, relies on the small size of the Bluetooth system and is easy to integrate, but Bluetooth signals are very susceptible to environmental interference, especially from human factors. To improve accuracy, a dense arrangement of Bluetooth LAN base stations is required, which leads to increased costs [4].

Ultrasonic indoor positioning technology relies on the penetration and immunity of ultrasound itself, and the easy availability of concentrated acoustic energy allows for very high accuracy in small-scale positioning. However, it is difficult to achieve large scale positioning due to multipath effects, non-visual errors, and rapid signal degradation [5]. Infrared technology is also highly accurate but has high environmental requirements, as the environment and heat and light sources can affect the accuracy of positioning, and it is far less effective in more obscured environments than in more open ones. It is also not easy to implement, as infrared systems are more complex to arrange and require multiple infrared generators [6]. Radio frequency technology, according to the use of different frequency bands of the electromagnetic wave can be applied to more detection conditions and environments. Compared with ultrasonic and visible light positioning, radio frequency identification technology has the advantage of being relatively unaffected by the line of sight environment, and at the same time, it has high positioning accuracy and a high transmission rate. However, since a large number of radio frequency transmitters need to be arranged, the cost is very high and the convenience is not high [7].

UWB technology is an emerging radio communication technology in recent years that calculates the location of a tag based on the geometric relationship between the base station and the tag through communication between the base station and known location information. UWB technology has the characteristics of good robustness, strong resistance to interference, strong resistance to multipath effect and accurate positioning compared to other technologies. However, UWB indoor positioning systems are more complex than other positioning systems and therefore more costly to implement [2, 7].

3. Geometric distance measurement methods

Estimating the location of the tag consists of two main steps; the first is to determine a model of the relative relationship between the tag and the base station, and the second is to build a system of equations based on the model to estimate the location using a positioning algorithm. Common models include TOA, AOA, RSSI and TDOA.

3.1. TOA

The Time-Of-Arrival (TOA) is the most common model in traditional ranging, which uses the time of flight of a wireless signal to estimate the distance between the base station and the tag. There are two common TOA methods, one measures the single time of flight of the signal and multiplies it by the speed of the signal flight to obtain the distance, while the other method measures the round-trip time of flight of the signal [8].

The One-Way time-of-flight Arrival based ranging method (OWR) is based on the principle shown in Figure 1. Under the condition that the clocks of the base station and the tag are synchronized, a tag sends a signal to its area at time, and a base station in the area receives the signal at time, so the distance between the tag and the base station is given by:

$$d_s = c \times (t_b - t_{aj}) \tag{1}$$

In which c represents the speed of light, and t_{bi} represents the time stamp when the i base station (i=1,2, 3...) received signal. t_{a_j} is the time stamp when the tag transmits the signal. The OWR ranging system is simple to implement and there is no limit to the number of tags with measurements. But everything has to be achieved with a high degree of synchronization between the base station and tag clocks, so it is difficult to implement in practice.

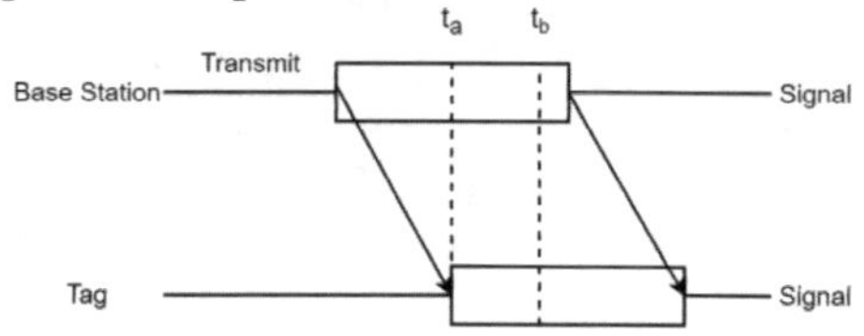

Figure 1. Principles of OWR.

The Two-Way time-of-flight Arrival (TWR) based ranging method is shown in Figure 2. The TWR method requires signal communication between the tag and the base station, unlike OWR, which uses a uniform device timestamp for its calculations to effectively avoid the problems associated with the inconsistent tag and base station clocks.

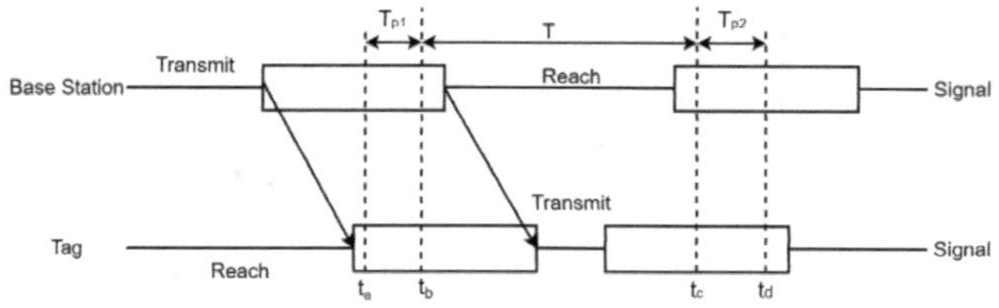

Figure 2. Principles of TWR.

In the process, the base station first transmits a ranging signal after a period of time t_0 after the tag i in t_b to receive the signal (i = 1, 2, 3, ...). After a period of time T, the tag in t_{ci} to send a response signal. The base station in t_{di} time to receive the signal. Use T_{P_1} to denote the time between the time the base station transmits the signal and the time the tag receives it, and T_{P_2} to denote the time the tag transmits the signal, and the base station receives it. Theoretically T_{P_1} should be equal to T_{P_2}, so we can get:

$$d_d = c \times T_{P_1} = c \times T_{P_2} \tag{2}$$

$$T_{P_1} = t_{b_1} - t_{a_1} \tag{3}$$

$$T_{P_2} = t_{d_1} - t_{c_1} \tag{4}$$

$$d_d = \frac{1}{2} \times c \times \left[(t_{b_1} - t_{a_1}) + (t_{d_1} - t_{c_1}) \right] \tag{5}$$

The TWR method solves the problems of synchronizing the tag and base station clocks on the one hand, and the number of tags on the other. As the base station can only communicate with one particular tag at a time, this limits the number of tags that can communicate at the same time.

From the principle of three-dimensional geometry, the distance from the tag to be measured to the base station is the radius, the base station is the center of the ball to make a ball, and different base stations for the center of the ball will intersect at a point. The location of the base station is usually used to locate four base stations. The above principle is shown in Figure 3:

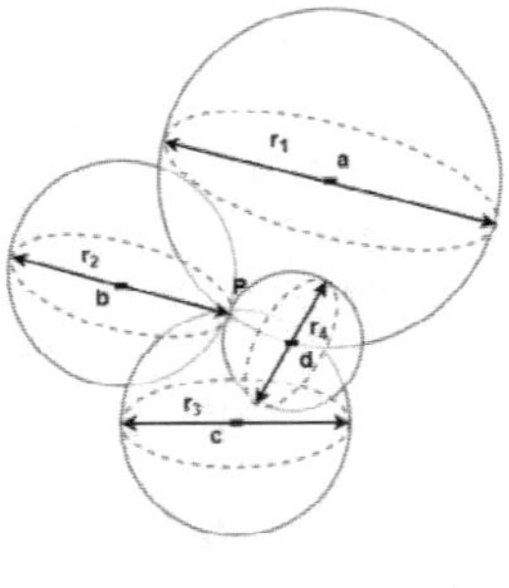

Figure 3. Geometric relations for TOA positioning.

As is shown in FIgure 3 that P is the locations of the tags to be measured, respectively, and the four base stations whose locations are known by their three-dimensional coordinates are:

$$\begin{cases} a(x_1, y_1, z_1) \\ b(x_2, y_2, z_2) \\ c(x_3, y_3, z_3) \\ d(x_4, y_4, z_4) \end{cases} \tag{6}$$

The distance between the four-base station to the tag is r_1, r_2, r_3, r_4. Let the three-dimensional coordinates of the base station be $p(x_0, y_0, z_0)$and we can get:

$$\begin{cases} r_1^2 = (x_1 - x_0)^2 + (y_1 - y_0)^2 + (z_1 - z_0)^2 \\ r_2^2 = (x_2 - x_0)^2 + (y_2 - y_0)^2 + (z_2 - z_0)^2 \\ r_3^2 = (x_3 - x_0)^2 + (y_3 - y_0)^2 + (z_3 - z_0)^2 \\ r_4^2 = (x_4 - x_0)^2 + (y_4 - y_0)^2 + (z_4 - z_0)^2 \end{cases} \tag{7}$$

Therefore, the following is a condensed formulation for the distance to the nth base station:

$$r_n = \sqrt{(x_n - x_0)^2 + (y_n - y_0)^2 + (z_n - z_0)^2} \tag{8}$$

3.2. TDOA

Like the time-of-arrival-based method TOA, Time Difference Of Arrival (TDOA) also uses the time of signal propagation in the air to achieve ranging. The difference is that instead of measuring the distance between a base station and the tag to be measured, TDOA needs to measure the distance difference between several different base stations to the tag distance to be measured to achieve the tag location. The principle is shown in Figure 4:

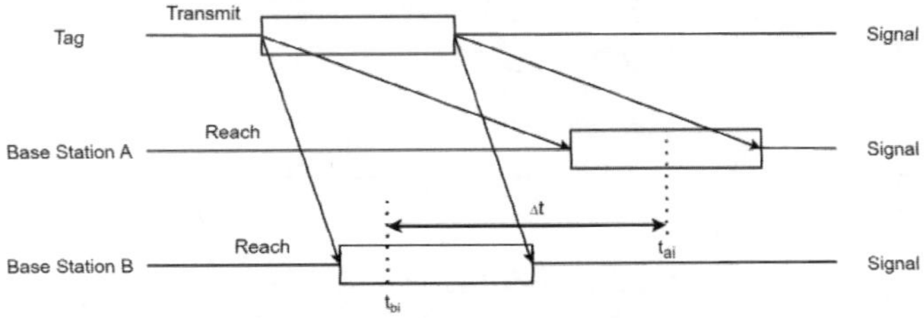

Figure 4. Principles of TDOA.

Use $d_a$and $d_b$to represent the distance from the base station A and B to the tag to be measured, c to represent the speed of light, and Δt to represent the time difference between the two base stations receiving the signal. So, the following equation for the distance difference can be obtained:

$$d_b - d_a = c \times \Delta t \tag{9}$$

The TDOA method does not require signal synchronisation between the tag and the base station, nor does it require the tag to communicate with the base station; it only requires the tag to send information to the base station. The TDOA method is shown in Figure 5:

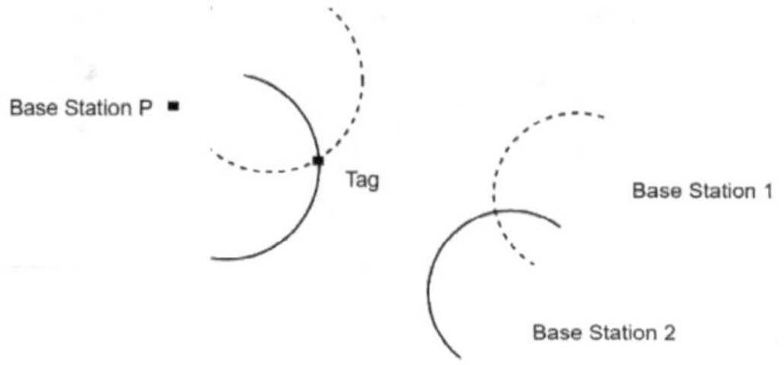

Figure 5. Geometric relations for TDOA positioning.

A base station P is selected as the main base station with coordinates (x_0, y_0), set the tag coordinates as (x, y) and the other base stations coordinates as (x_i, y_i). The above coordinates should satisfy the following equation:

$$r_{0i} = r_i - r_o = \sqrt{(x_i - x)^2 + (y_i - y)^2} - \sqrt{(x_0 - x)^2 + (y_0 - y)^2} \quad (10)$$

The distance between the first base station and the tag to be measured and the distance difference between the base station and the tag to be measured are denoted by. The principle of the model is: the distance difference between the tag to be measured and the different base stations has been obtained, usually a base station is selected and the distance difference between the second base station and that base station to the same tag is found, respectively. If the distance difference between two base stations and the base station is measured, two sets of hyperbolas can be obtained, and the intersection of the two sets of hyperbolas is the location of the tag to be measured [8-9].

3.3. AOA

When using the Angle Of Arrival (AOA) positioning system, it is typically necessary to set up the antenna array for angle identification, ranging in accordance with the various signals, and the horizontal axis formed by the angle of incidence reverse extension so that they intersect at a point, the point being the location of the tag to be measured, according to the specific principle illustrated in Figure 6:

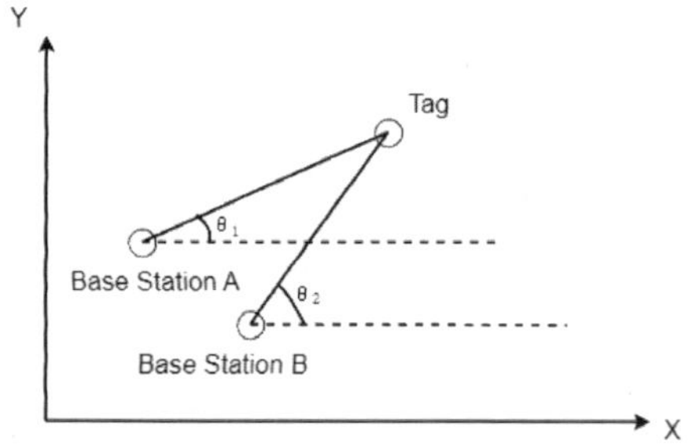

Figure 6. Principles of AOA.

The coordinates of the base station are (x_1, y_1),the coordinates of the base station are (x_2, y_2), let the coordinates of the tag be (x, y), the three should satisfy the following relationship equation:

$$\begin{cases} \tan\theta_1 x - y = \tan\theta_1 x_1 - y_1 \\ \tan\theta_2 x - y = \tan\theta_2 x_2 - y_2 \end{cases} \quad (11)$$

So, the N base station (x_n, y_n) should follow:

$$\tan\theta_n x - y = \tan\theta_n x_n - y_{2n} \quad (12)$$

There are N(N>2) base stations in the area. The incidence angle of each tag to the base station is θ_n, the coordinate of each base station is (x_n, y_n), according to the above relationship for the coordinates of the tag to be measured. The AOA method also avoids the problem of unsynchronized tag and base station clocks and is easy to implement, but it requires a large number of antenna matrices to be arranged and the cost is relatively high.

3.4. RSSI

The RSSI algorithm is based on the idea that a signal gets weaker as it moves through space. Therefore, the following equation holds true for the power received by the tag and the signal's distance:

$$P_b(S) = \frac{P_i G_j G_{bz} \lambda^{\tau}}{(4, \pi s)^{\tau}} \tag{13}$$

Where P_b is the frequency of the signal received by the tag, P_i is the transmit power of the signal transmitted by the base station, S is the distance transmitted, G_j is the gain when the base station transmits the signal, G_{bz} is the signal gain when the tag receives the signal, λ is the wavelength of the signal, and τ is the regional environmental path loss factor.

The RSSI method also does not require a high degree of synchronization between the base station and tag clocks because it is independent of time, but the loss and loss factor of the signal during propagation are greatly influenced by the environment, and subtle changes in the environment can lead to large changes in the results, so the error in the data is larger, and the resulting positioning error is larger [10].

3.5. Double-sided Two-way Ranging

The accuracy of ranging using various signal bandwidths varies, however UWB positioning systems using TOA or TDOA methods are substantially superior to AOA and RSSI methods because UWB has a very wide frequency domain signal with strong temporal resolution [11].

The TOA method only requires the measurement of the time stamp at each point in the process, after which the transmission time TOF of the signal is used for ranging. However, a TOF signal measurement time mistake of 1ns can result in a 30cm or even greater distance error. This is caused by system clock jitter due to crystal drift. It is therefore necessary to ensure that the clocks of the tag and the base station are highly synchronized. This is often difficult to achieve the desired synchronization in a practical environment or adds significantly to the complexity of the system, so a direct calculation of TOF can lead to large measurement errors. The bilateral bi-directional ranging method uses TOA as the basis for optimization and improves accuracy without the need to ensure that the tag and base station clocks are highly synchronized [11].

The bilateral bi-directional ranging method requires separate communication between the tag and the base station, i.e., one bilateral bi-directional ranging with all base stations in one cycle, and is shown in Figure 7:

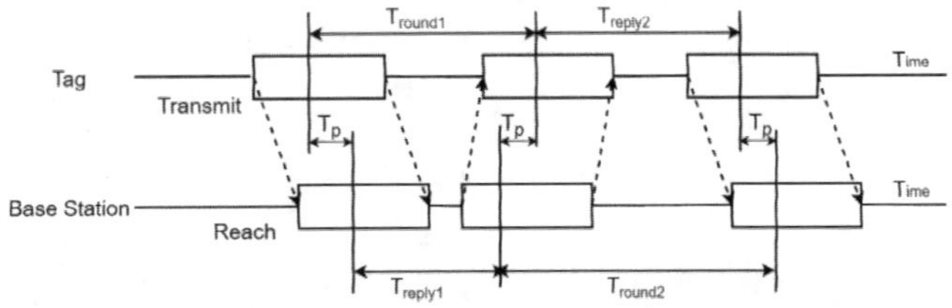

Figure 7. Principle of Double-sided Two-way Ranging.

The noise of the DS_TWR algorithm comes mainly from the timestamp accuracy of the RF chip, non-data effects, and the effects caused by multipath effects. The error introduced by its clocking can be expressed as follows:

$$\Phi = T_p \times \left(1 - \frac{k_1 + k_2}{2}\right) \tag{14}$$

Where k_1 and k_2 denote the ratio of the actual frequency to the theoretical maximum frequency of the base station clock and tag clock respectively, both of which should be between 0.99998 and

1.00008. Assuming that the clock accuracy of the tag and the base station is poor, both are 10ppm level, let the distance between the base station and the tag be 10m, based on the speed of light as 3×10^8m/s to calculate, T_p should be 33.333ns, so the maximum value of Φ obtained according to the maximum value of k_1 and k_2 is 0.2mm, which can be neglected.

4. The development of UWB

4.1. Advantages and disadvantages of UWB

After 20 years of development, UWB has now received a lot of attention. From the point of view of current developments, UWB has many advantages: Firstly, UWB has high positioning accuracy. Secondly, UWB positioning has very low positioning latency. Thirdly, the UWB positioning technology is highly resistant to multipath effects. Finally, UWB system has high system robustness [7].

However, UWB indoor positioning technology is currently facing many difficulties:

Firstly, the price of positioning base stations and tags for UWB systems is significantly higher than the price of tags and base stations for other positioning systems. The quantity of base stations has an impact on positioning precision; hence, striving to increase accuracy also raises expenses. The quantity of base stations has an impact on positioning precision, hence striving to increase accuracy also raises expenses [12].

Secondly, it has high research and development difficulties. Most of the current UWB positioning systems use the DW1000 chip, and there are currently fewer algorithms based on the DW1000 chip. Problems such as positioning errors are dependent on algorithm optimization, which can greatly increase costs [7].

4.2. Recent advances in UWB

Recent developments in UWB have focused on how to suppress non-line-of-sight errors, improve algorithm accuracy, and increase positioning accuracy while controlling system complexity and cost.

The biggest issue now affecting UWB placement is the main system's low theoretical cost but high actual cost. According to known geometric positioning methods, reducing the number of base stations, i.e., anchor points, can reduce the computational effort of the algorithm and can reduce the problem of timestamp synchronization in the TOA method. The use of a single anchor point reduces the cost of the hardware on the one hand and the software algorithm on the other.[7, 13]

In addition to the higher cost, UWB positioning is susceptible to non-city errors. A TOA-based approach using Kalman filtering to suppress non-line-of-sight errors was used to improve the accuracy of 3D TOA positioning by Dongchen Ni et al. a new pedestrian heading projection (PDR) assisted UWB indoor positioning was proposed by Ming Xia et al.[9]. The researchers used the distance difference between UWB and PDR to calculate an accuracy factor model for the non-visual range error. This accuracy factor model was used to determine whether the UWB could be accurately positioned, and if this threshold was exceeded, the PDR was used instead of the UWB. The researcher's experiments showed that this method could significantly improve the positioning accuracy of the UWB. [14-16]

Unlike Josef Kulmer et al. who used a single anchor point Alwin Poulose et al. chose to use multiple anchor points [12]. Alwin Poulose et al. propose an indoor position estimation device using the extended Kalman filter (EKF) [2]. In their simulations they use methods such as least squares estimation for comparison and the results shows that the use of the extended Kalman filter (EKF) with aided indoor positioning systems can help improve the accuracy of UWB indoor positioning while reducing the complexity of the system for the same number of anchor points.

5. Conclusion

This paper summarizes the current common indoor positioning methods and compares their advantages and disadvantages. It is noted that the paper focuses on common geometric positioning methods for UWB and their recent development. The dominant geometric positioning method for UWB is now well established and will not change significantly with minor optimization. Optimization of UWB positioning should be based around suppressing non-visual errors and optimizing the positioning algorithm around the DW1000 chip. Judging from the speed and interest in UWB, UWB indoor positioning has a high scope for development and potential economic benefits.

References

[1] Alarifi, A., Al-Salman, A., Alsaleh, M., Alnafessah, A., Al-Hadhrami, S., Al-Ammar, M. A., & Al-Khalifa, H. S. (2016). Ultra-wideband indoor positioning technologies: Analysis and recent advances. Sensors, 16(5), 707.

[2] Poulose, A., Emeršič, Ž., Eyobu, O. S., & Han, D. S. (2020, October). An accurate indoor user position estimator for multiple anchor uwb localization. In 2020 international conference on information and communication technology convergence (ICTC)(pp. 478-482). IEEE.

[3] Shang, S., & Wang, L. (2022). Overview of WiFi fingerprinting-based indoor positioning. IET Communications, 16(7), 725-733.

[4] Bekkelien, A., Deriaz, M., & Marchand-Maillet, S. (2012). Bluetooth indoor positioning. Master's thesis, University of Geneva.

[5] Ijaz, F., Yang, H. K., Ahmad, A. W., & Lee, C. (2013, January). Indoor positioning: A review of indoor ultrasonic positioning systems. In 2013 15th International Conference on Advanced Communications Technology (ICACT)(pp. 1146-1150). IEEE.

[6] Lee, C., Chang, Y., Park, G., Ryu, J., Jeong, S. G., Park, S., ... & Lee, M. H. (2004, November). Indoor positioning system based on incident angles of infrared emitters. In 30th Annual Conference of IEEE Industrial Electronics Society, 2004. IECON 2004(Vol. 3, pp. 2218-2222). IEEE.

[7] Sakpere, W., Adeyeye-Oshin, M., & Mlitwa, N. B. (2017). A state-of-the-art survey of indoor positioning and navigation systems and technologies. South African Computer Journal, 29(3), 145-197.

[8] Mahfouz, M. R., Fathy, A. E., Kuhn, M. J., & Wang, Y. (2009, September). Recent trends and advances in UWB positioning. In 2009 IEEE MTT-S International Microwave Workshop on Wireless Sensing, Local Positioning, and RFID(pp. 1-4). IEEE.

[9] Ni, D., Postolache, O. A., Mi, C., Zhong, M., & Wang, Y. (2019, March). UWB indoor positioning application based on Kalman filter and 3-D TOA localization algorithm. In 2019 11th International Symposium on Advanced Topics in Electrical Engineering (ATEE)(pp. 1-6). IEEE.

[10] García, E., Poudereux, P., Hernández, Á., Ureña, J., & Gualda, D. (2015, March). A robust UWB indoor positioning system for highly complex environments. In 2015 IEEE International Conference on Industrial Technology (ICIT)(pp. 3386-3391). IEEE.

[11] Kwak, M., & Chong, J. (2010, September). A new double two-way ranging algorithm for ranging system. In 2010 2nd IEEE InternationalConference on Network Infrastructure and Digital Content (pp. 470-473). IEEE.

[12] Kulmer, J., Hinteregger, S., Grosswindhager, B., Rath, M., Bakr, M. S., Leitinger, E., & Witrisal, K. (2017, May). Using DecaWave UWB transceivers for high-accuracy multipath-assisted indoor positioning. In 2017 IEEE International Conference on Communications Workshops (ICC Workshops)(pp. 1239-1245). IEEE.

[13] Shen, Y. (2022, December). Applications of UWB Positioning Technology in Smart Home. In 2022 6th International Seminar on Education, Management and Social Sciences (ISEMSS 2022)(pp. 874-883). Atlantis Press.

[14] Alarifi, A., Al-Salman, A., Alsaleh, M., Alnafessah, A., Al-Hadhrami, S., Al-Ammar, M. A., & Al-Khalifa, H. S. (2016). Ultra-wideband indoor positioning technologies: Analysis and recent advances. Sensors, 16(5), 707.

[15] Minne, K., Macoir, N., Rossey, J., Van den Brande, Q., Lemey, S., Hoebeke, J., & De Poorter,

E. (2019). Experimental evaluation of UWB indoor positioning for indoor track cycling. Sensors, 19(9), 2041.
[16] Xia, M., Xiu, C., Yang, D., & Wang, L. (2018, March). A novel PDR aided UWB indoor positioning method. In 2018 Ubiquitous Positioning, Indoor Navigation and Location-Based Services (UPINLBS)(pp. 1-7). IEEE.

Research and mechanism of iron-based superconductors

Cheng Bi
University College London, UK

dannyxu43@gmail.com

Abstract. The discovery of new iron-based superconductor $LaFeAsO_{1-x}F_x$ with Tc of 26 K in 2008 had quickly revived interest in the exploration of iron-based superconductors. After several years of research on this superconducting material, iron-based superconductors have become another family of high-temperature superconductors after copper-based superconductors. The crystal structure of LaOFeAs is P4/nmm family, and by doping with fluorine, $LaO_{1-x}F_xFeAs$ exhibits superconductivity because both structural distortion and magnetic order are suppressed. In this review, the development and experimental results of such iron-based high Tc materials will be presented. We describe the synthetic method of this class of superconductors, the physical properties of the precursors, the doping methods that can produce superconductivity, the pressure effect, and the prospects of this new iron-based high Tc family.

Keywords: superconductivity, iron-based superconductors, schematic phase diagrams.

1. Introduction

Superconductivity is the most widely studied phenomena in the condensed matter physics field in recent years. Finding out the internal mechanism of superconducting electron pairs and exploring a kind of material having high critical temperature (Tc) are the most significant targets in this field now. In 2008, Hideo and his coworkers discovered a new kind of iron-based superconductor $LaFeAsO_{1-x}F_x$ with Tc of 26K, which had drawn much interest in researching iron-based superconductors [1]. After that, worldwide researchers worked hard on iron-based superconductors and hundreds of new materials of this family of superconductors have occurred so far. Therefore, iron-based superconductors became the second members of high-Tc family after cuprates [2-5]. Traditionally, the antagonistic relationship between magnetism and superconductivity made scientists doubt the magnetic elements, especially ferromagnetic elements such as iron [1,3]. Fe-based superconductors (FeSCs) have challenged the theories of high-Tc superconductivity strongly and changes in the understanding of superconductivity theory are needed [6]. The basic nature of different FeSCs is reported to be similar: chemical substitution plays an important role. More ionic character atomic bonds are formed within the spacer layers. At zero doping, spin density wave (SDW) instability comes first; at a finite doping, s+ superconducting instability occurs at a higher temperature [7].

Iron-based superconductors (IBSCs) have some specific properties like good grain boundary, robustness to impurity and high upper critical field [8]. This article focuses on some important iron-based layered systems and provides an overview of high critical temperature FeSCs superconductors experimental results. Some important theoretical perspectives are reported to update understanding of the properties of superconductivity.

2. $LnO_{1-x}F_xFeAs$ superconductors (Ln = lanthanides)

2.1. Review of $LaO_{1-x}F_xFeAs$ superconductors

$RO_{1-x}F_xFeAs$ materials are the first non-copper superconductors whose Tc is over 50K and then draw much interest in the superconductivity area [1, 9, 10]. The resistivity of its parent compounds, LaOFeAs, has a significant drop at 150K [9, 11] and LaOFeAs is regarded to has an antiferromagnetic SDW instability below 150K. Superconductivity may arise by suppressing the antiferromagnetic SDW. Clarina de la Cruz and his colleagues tried to research the magnetic and structural order in non-superconducting LaOFeAs and superconducting $LaO_{0.92}F_{0.08}FeAs$ by applying neutron diffraction [12]. The crystal structure of LaOFeAs is group P4/nmm. When cooling to 4 K, there was a structural phase transition according to the BT-1 diffractometer and the same transition is suppressed in the $LaO_{0.92}F_{0.08}FeAs$ system [12]. The sample turned from tetragonal to monoclinic structure when cooling to a specific temperature. Such phase transition is related to the specific heat anomalies [11] and resistivity [1]. They found that the structural distortion happened before the magnetic order establishment: long-range SDW antiferromagnetic order was developed at 137K [12]. This kind of superconductivity has a simple stripe-type antiferromagnetic structure [12]. By doping with fluorine, $LaO_{1-x}F_xFeAs$ showed superconductivity because both the structure distortion and magnetic order were suppressed.

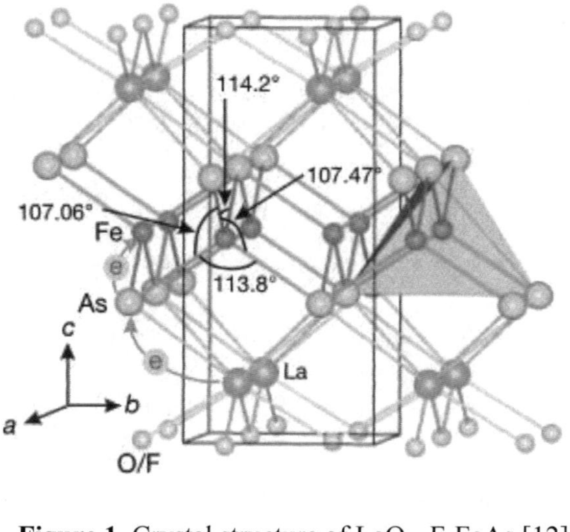

Figure 1. Crystal structure of $LaO_{1-x}F_xFeAs$ [12].

Hiroki Takahashi et al. reported the relationship between pressure and superconductivity of $LaO_{1-x}F_xFeAs$ materials [13]. They believed that the La-O chemical bond in the LaO layer in the layered structure of superconductor LaOFeAs is ionic and positively charged, while the Fe-As is mainly covalent and negatively charged. The chemical formula can be written as $(La^{3+}O^{2-})^+(FeAs)^-$ [13]. The conduction carriers are confined within the layers, which leads to strong interactions among the electrons [1]. When the pressure increased to above 3 GPa, the critical temperature decreased sharply and at 4 GPa the Tc reached to maximum of 43 K. After doping with fluoride ion, both a and c axes shrink and the anisotropic shrinkage was induced by external pressure [14]. And because of doping with fluoride ion, an extra positive charge is added into the insulating layer, while a negative charge is provided in the conductive layers. This enables the tuning of interlayer interactions through changes in carrier density without causing severe crystal defects [1].

2.2. $LnO_{1-x}F_xFeAs$ material synthesis methods

Different methods have been applied to the iron-based superconductors $LnO_{1-x}F_xFeAs$ preparation. Solid state method was first applied by Kamihara et al. and then was widely used in the iron-based superconductors preparation [1]. LnAs, LnF_3, Fe_2As, FeAs and rare earth oxides are taken as raw materials to synthesize $LnO_{1-x}F_xFeAs$ [15]. The $LnO_{1-x}F_xFeAs$ materials were synthesized by stoichiometric mix of those raw materials and sintering at 1423 to 1523 K for 40-60 hours [1, 12, 13]. And the samples were ground and sintered at a higher temperature (1573 K) in order to improve the densification [13, 14]. And Chen and his colleagues changed rare earth oxide to Fe_2O_3 to take advantage of its high stability [15]. There is a more efficient method to prepare $LnO_{1-x}F_xFeAs$ materials at high temperatures, high pressure method (HP). Using HP method can avoid fluorine loss in the vacuum quartz tube. The raw material pellets in boron nitride crucibles were sintered at 1523K and 6 GPa for 2 hours. However, since the sintering time was quite short, there were some impurity phases like rare earth oxides. Yang and his colleagues found that the Sm based sample prepared by HP method had higher Tc and smaller lattice parameters by eliminating the volatilization of fluoride ion. Using HP method can make some heavy rare earth elements (Dy, Tb) LnOFeAs system superconductors of which the Tc is over 40 K [15, 16]. Jia et al. reported another method which can tailor the intrinsic properties of Fe based superconductors [17]. The flux method was applied to form single crystal. They used NaCl/KCl as the flux and sealed the raw materials into the vacuum quartz tube. They were heated at 1423 K for 10 days and then synthesized to single phase $NdFeAsO_{0.82}F_{0.18}$. Flux method takes more time to prepare the products, but it can effectively reduce the processing temperature and facilitate the growth of high-quality single crystal materials [17].

3. Parent materials and doping element

After the report of new iron-based $LaO_{1-x}F_xFeAs$ superconductors, many superconducting materials were reported in the past few years. Researchers used different rare-earth elements to replace the La and change the crystal constants. Some new iron-based superconductors had an important breakthrough and the Tc became above 50K such as $SmO_{1-x}F_xFeAs$, $PrO_{1-x}F_xFeAs$, $NdO_{1-x}F_xFeAs$ and $CeO_{1-x}F_xFeAs$ [14, 15, 18]. In subsequent studies, it was found that the 1111-type superconductor had the highest initial Tc, which were about 55-57 K while Tc of other types of iron-based superconductors were below 40K [19]. Those materials are tetragonal symmetry and Fe plays the central part of the superconductivity. The Pauli para metals would undergo a crystallographic transition to monoclinic/orthorhombic when temperatures become very low [20]. When the anti-ferromagnetism vanishes through loading external/chemical pressure or doping carriers, superconductivity will emerge.

The parent phase exhibits superconductivity at low temperatures, but the Tc value of this material is quite low, such as Tc=4 K for LaOFeP [1]. There is a strong relationship between the magnetic ordering in the parent phase and the resulting Tc. High Tc iron-based superconductors can be obtained by applying carrier doping to eliminate the long-range spin order in layered-structure iron phosphides. F-doped LaOFeAs IBSCs show a crystalline transition when lowered to a specific temperature. The lattice decreases systematically with the increasing concentration of dopant, suggesting chemical pressure inside the material. The thickness of the layered structure decreases after doping fluoride ion, indicating that the doping enhances Coulomb and polarization interactions between layers [21]. And the SDW anomaly peak was suppressed by doping F, which makes the Ts lower, thereby switching to the superconducting transition under optimal doping. Figure 2. Shows different "1111" compounds of lanthanides (La, Ce, Sm, Nd and Gd) with optimal fluorine doped [22]. The ρ-T graph shows an anomaly around 150K associated with the SDW state, below which the resistivity drops sharply linearly with temperature [22].

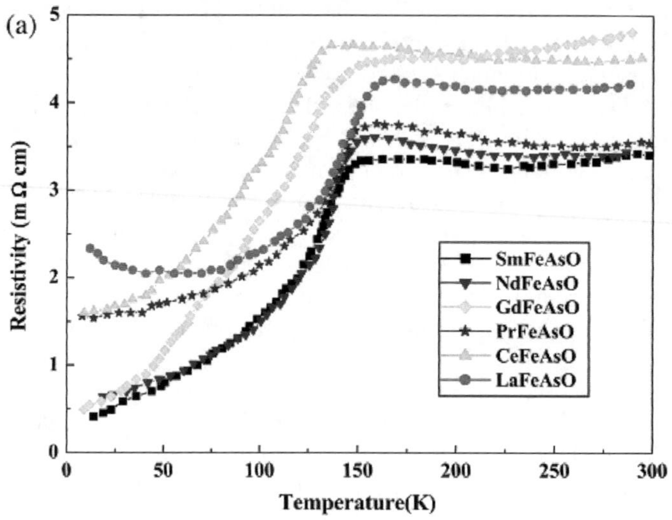

Figure 2. Temperature dependence of resistivity in LnOFeAs [22].

The structure of "1111" material is 2 dimensional with alternant insulating LaO layers and conductive FeAs layers [23]. Because of the energy shift, the generated electrons are transferred to the FeAs layers after doping fluoride ions. As shown in Figure 3, when the anti-ferromagnetism disappears, the critical temperature appears in 1111 system [20]. Ren et al. used the Ln (rare earth metal) atoms with smaller ionic radius to replace the La atom and the result was satisfying: Tc increased from 26 K to 55 K [18].

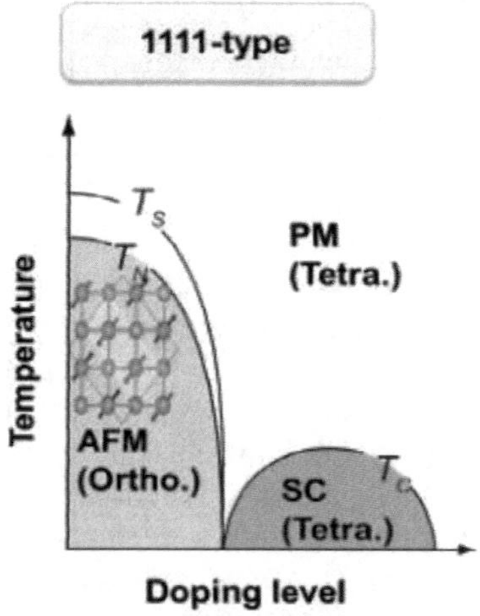

Figure 3. Schematic phase diagrams of "1111" materials [20].

Because fluoride ion has relatively low solubility at the oxygen site, the doping levels were insufficient and led to unsatisfactory results. S. Matsuishi et al. used hydrogen ion rather than fluoride ion for doping [24]. They used high-pressure method to synthesize the H-substituted rare earth element "1111" system compounds, $CeFeAsO_{1-x}H_x$ [24]. They found that as the size of rare earth metal ions decreased the doping level declined. As shown in Figure 4, "1111" system LaFeAsO had 2-done structure: one was the same as that of $LaFeAsO_{1-x}F_x$, the other, however, had higher critical temperature (36 K). For synthetic materials of other rare earth elements, the system has only one dome, but its range is much wider than that of fluoride ion alternative materials for the same lanthanide ion system [24, 25].

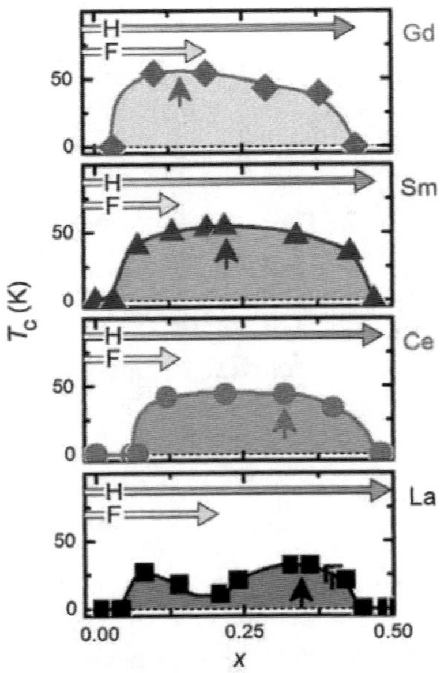

Figure 4. Using hydrogen ion to realize wider superconducting rages than fluoride ion. Tc of different LnFeAsO (Gd, Sm, Ce, La) materials [24].

Researchers also tried hole doping in LaFeAsO system. Wen et al. reported a hole doping superconductor with Tc = 25 K, $La_{1-x}Sr_xFeAsO$ (when x = 0.13) [26]. They introduced holes into LaFeAsO system and induced superconductivity through partially replacing La with Sr [26]. The hole charge carrier mechanism was proved by Hall effect measurements. And the thorium was also used to substitute the rare earth ion sites to increase the Tc. Wang and his colleagues synthesized $Gd_{1-x}Th_xFeAsO$ which had a Tc of 56 K [27]. The Tc was much higher than that of fluoride ion doped material.

4. Conclusion

In this review, different types of iron-based are summarized. Iron-based superconductors have become another family of high-temperature superconductor after copper-based superconductors since a new kind of iron-based superconductor $LaFeAsO_{1-x}F_x$ with Tc of 26K was discovered by Hideo et al. in 2008. One the basis of LaOFeAs, by replacing lanthanide elements, doping fluorine ions, doping metal ions of different valence states and so on to increase the Tc. We also introduced various of synthetic methods for some LaOFeAs system superconductors. With the effort of many researchers, several kinds of iron-based superconductors with Tc exceeding 50 K have appeared. It was found that in LaOFeAs system superconductors the 1111-type superconductor had the highest initial Tc, which were about 55-57 K. The "1111" type material has Fe-As layer crystal structure. When the anti-ferromagnetism vanishes through loading external/chemical pressure or doping carriers, superconductivity will emerge. The development of iron-based superconductors has introduced new high Tc family into the field of superconductors, and it has been found that high-Tc superconductors are more common in iron-based superconductors. Therefore, iron-based superconductors may exist with various compounds having Tc values higher than those found in cuprates. In short, iron-based superconductors have quite good prospects.

References

[1] Kamihara Y, Watanabe T, Hirano M, and Hosono H. 2008. Iron-based layered superconductor $La[O_{1-x}F_x]FeAs$(x=0.05–0.12) with T c =26K. *J. Am. Chem. Soc* 130 3296–97.

[2] Athena S. Sefat et al. 2008. Electronic correlations in the superconductor $LaFeAsO_{0.89}F_{0.11}$ with low carrier density. *Phys. Rev. B* 77 174503.

[3] Shekhar C. et al. 2008. Synthesis and microstructural studies of iron oxypnictide $LaO_{1-x}F_xFeAs$ superconductors. *Supercond. Sci. Technol.* 22 015005.

[4] Tropeano M. et al. 2008. Thermal properties of $SmFeAsO_{1-x}F_x$ as a probe of the interplay between electrons and phonons. *Phys. Rev. B* 78 094518.
[5] Jaroszynski J. et al. 2008. Comparative high-field magnetotransport of the oxypnictide superconductors RFeAs$O_{1-x}F_x$ (R=La, Nd) and $SmFeAsO_{1-\delta}$. *Phys. Rev. B* 78 064511.
[6] Paglione J, Greene R. 2010. High-temperature superconductivity in iron-based materials. *Nature Phys* 6 645–58.
[7] M. G. Vavilov et al. 2010. Coexistence between superconducting and spin density wave states n iron-based superconductors: Ginzburg-Landau analysis. *Supercond. Sci. Technol* 23 054011.
[8] Akiyasu Y. et al. 2018. Recent advances in iron-based superconductors toward applications. *Materials Today* 21 278-302.
[9] Chen G. F. et al. 2008. Superconducting properties of Fe-basedlayered superconductor $LaO_{0.9}Fe_{0.1-d}$ FeAs. *Phys. Resv. Lett* 101 057007.
[10] Yen Z. A. et al. 2008. Superconductivity at 52 K in iron-based F-doped layered quaternary compound $Pr[O_{1-x}F_x]FeAs$. *Mater. Res. Innov* 12 105-06.
[11] Dong J. et al. 2008. competing orders and spin-density-wave instability in $La(O_{1-x}Fe_x)FeAs$. Preprint at Æhttp://arxiv.org/abs/0803.3426v1æ.
[12] De la Cruz C, Huang Q, Lynn J. et al. 2008. Magnetic order close to superconductivity in the iron-based layered $LaO_{1-x}F_x$ FeAs systems. *Nature* 453 899–902
[13] Takahashi H, Igawa K, Arii K. et al. 2008. Superconductivity at 43 K in an iron-based layered compound $LaO_{1-x}F_xFeAs$. *Nature* 453 376–78.
[14] Chen X, Wu T, Wu G. et al. 2008. Superconductivity at 43 K in $SmFeAsO_{1-x}F_x$. *Nature* 453 761–62.
[15] Chen G F, Li Z. et al. 2008. Superconductivity at 41 K and its competition with spin-density-wave instability in layered $CeO_{1-x}F_xFeAs$. *Phys. Rev. Lett* 100 24700.
[16] J W Bos, G S Penny. et al. 2008. High pressure synthesis of late rare earth RFeAs(O,F) superconductors; R=Tb and Dy. *Chem. Commum* 31 3634.
[17] Y Jia, P Cheng. et al. 2008. Critical fields and anisotropy of $NdFeAsO_{0.82}F_{0.18}$. *Appl. Phys. Lett* 93 032503.
[18] Z A Ren, J Yang, W Lu. et al. 2008. Superconductivity at 55 K in iron-based F-doped layered quaternary compound $SmO_{1-x}F_xFeAs$. *Mater. Res. Innovations* 12 105.
[19] C Wang, L Li. 2008. et al. Thorium-doping induced superconductivity up to 56 K in $Gd_{1-x}Th_xFeAsO$. *Europhys. Lett* 83 67006.
[20] H Hosono, K Kuroki. 2015. Iron-based superconductors: Current status of materials and pairing mechanism. *Physica C* 514 399-422.
[21] T Nomura, S W Kim. 2008. et al. Crystallographic phase transition and high-Tc superconductivity in LaFeAsO:F. *Supercond. Sci. Technol* 21 125028.
[22] Z A Ren, Z X Zhao. 2009. Research and prospects of iron-based superconductors. *Adv. Mater* 21 4584-4592.
[23] S Margadonna, Y Takabayashi. et al. 2009. Pressure evolution of the low-temperature crystal structure and bonding of the superconductor FeSe (Tc=37 K). *Phys. Rev. B* 80 064506
[24] S Matsuishi, T Haanna. et al. 2012. Structural analysis and superconductivity of $CeFeAsO_{1-x}H_x$. *Phys. Rev. B* 85 014514.
[25] S Limure, M Hiraishi. et al. 2014. Bipartite magnetic parent phases in the iron oxypnictide superconductor. *Nature Phys* 10 300-03.
[26] H Wen, G Mu. et al. 2008. Superconductivity at 25 K in hole-doped $(La_{1-x}Sr_x)OFeAs$. *Europhys Lett* 82 17009.
[27] C Wang, L Li. et al. 2008. Thorium-doping-induced superconductivity up to 56 K in $Gd_{1-x}Th_xFeAsO$. *Europhys. Lett* 83 67006.

Overview of gravitational waves: Formation, polarization and detection

Jing Zhou

Department of Physics, Fudan University, Shanghai, 200433, China

zdlzsj@163.com

Abstract. Gravitational waves (GWs) are reliable for proving general relativities. Also, they can be used in detecting physics phenomena in the universe. Motions of bodies make spacetime curved, and then gravitational waves are formed. Additionally, GWs can be produced in the early universe. A mathematical description of gravitational waves derived using linearized theory and transverse-traceless gauge shows GWs polarizations and the relationship between perturbation and strain. GW signal was first detected in the binary system PSR1913+16. Data presented an orbital decay, which followed general relativity's prediction. Direct detections today, designed according to GWs properties, are carried out by resonant detectors, interferometers, and pulsar timing arrays. Noise reduction is a key to accurate detection. Furthermore, analyzing the B-mode cosmic microwave background can be used in detecting primordial gravitational waves.

Keywords: Formation Of GW, GW Polarization, PSR 1913+16, Detectors, CMB Polarization.

1. Introduction

Gravitational waves are associated with the curvature of spacetime. They predict general relativity and any other relativistic theories of gravity. They provide a new method to examine general relativity and its alternatives in the high-speed, robust field regime [1]. By studying gravitational waves, we can get a picture of the coalescence of compact binaries.

We are looking forward to finding new physics corresponding to the connecting part of two inspiral bodies. Also, we can get information about the Big Bang and the early universe from gravitational waves formed around inflation.

This paper introduces general ideas of gravitational waves. Section 2 is about the formation of gravitational waves. GWs can be produced around the time of inflation. They can also be generated from the motions of bodies. Section 3 shows the polarization of gravitational waves. Assumptions and mathematical derivation of polarization are present. Also, strain, a property of gravitational waves related to their polarization, is mentioned in section 3. Section 4 is about indirect detection. The signal from PSR 1913+16 shows orbital period decays, which is consistent with general relativity's prediction. Section 5 presents three gravitational wave detectors: resonant, interferometer, and pulsar timing arrays. Operating principles and suitable frequency for each sensor are shown in this section. What's more, it will contain an error analysis. Section 6 briefly introduces using CMB (cosmic microwave background) polarization to detect primordial gravitational waves.

2. Formation of Gravitational Waves

2.1. Motion of Bodies

The motion of massive bodies like compact binaries can produce gravitational waves. Activities of bodies make spacetime curved. Gravitational waves are "ripples" in the curvature of spacetime and propagate at the speed of light. Most GWs caused by motions have frequencies ranging from $10^{-4}Hz$ to $10^{4}Hz$, depending on the masses of bodies.

2.1.1. Coalescence of Compact Binaries. Bodies in compact binary systems are neutron stars or black holes. The coalescence of binaries can be divided into 3 phases: inspiral, merger, and ring-down. Gravitational waves are produced in all three stages. During inspiral, bodies in a binary system orbit around each other. Distance between two bodies shortens, and relative velocity increases. The orbital period is reduced, and the system emits orbital energy through gravitational waves. Two bodies merge into a single one the merger, forming more intensive gravitational waves. Then, the strength of gravitational waves decays, and the system goes into the ring-down phase. During this stage, the new body emits gravitational waves to achieve equilibrium [2].

As for binary black holes, strain (Strain can reflect the strength of gravitational waves. More details are shown in section 3.) of gravitational waves is mildly increased during the inspiral. And then, strain changes rapidly in the merger phase. During ring-down, strain oscillates even milder and rests in the end. FIG.2, in reference, shows the strain's change of coalescing binary black holes. Themes of GWs from other compact binaries change in the same tendency with different values [3].

2.1.2. Stellar Core Collapse. The core collapse in massive stars is a promising source of gravitational waves. It contains necessities for generating gravitational waves: a mass of $1 - 100M_{\odot}$ flows in a compact region at relativistic speed. Instabilities caused by highly rapid rotation (Most massive-star models indicate that such a process is unlikely to exist.) in core or asymmetric dynamics of core-collapse can lead to the emission of gravitational waves [4].

2.1.3. Rotating Neutron Stars. A rotating neutron star or GW pulsar is a prototypical source for continuous gravitational waves. One mechanism concerns the toroidal magnetic field B_t, which distorts neutron stars into a prolate shape. An elastic neutron star is unstable under a large B_t , and the star's magnetic axis mismatch with the star's angular momentum. Gravitational waves are more likely to be emitted when the magnetic axis and angular momentum are orthogonal [5].

2.2. In Early Universe

One model of generating gravitational waves is related to fluctuations. During inflation, initial quantum fluctuations were amplified into classical perturbation outside the visible universe. This process formed tensor modes, which can be viewed as gravitational waves. Gravitational waves generated in this way have an extensive frequency range [6].

Another model of gravity-wave production is associated with inhomogeneities. In many inflation models, accelerated expansion ends with preheating. During preheating, energy was extracted from the inflation field non-perturbatively and non-thermally, leading to a non-thermal distribution. Preheating introduces large and transient density inhomogeneities, which can be a source of stochastic gravitational waves [7]. In this model, the frequencies of gravitational waves are in the Hz to GHz range [8].

3. Polarization

As a ripple in the curvature of spacetime, the passage of gravitational waves leads to changes in the distance of particles in different directions, which present polarizations of gravitational waves. There are two types of polarizations: "plus" and "cross". If a gravitational wave propagates along z direction, there are six possible polarization modes. However, only 2 of them are allowed in general relativity. Fig.5.8 in reference shows the polarization modes of GWs [9].

3.1. Approximation in General Relativity

There is no apparent difference between the metric representing waves and other metrics. Hence, three approximations-linearized theory, short-wave approximation, and post-Newtonian theory-are applied to make a distinction.

Linearized theory approximates under a weak field, where motions happen in nearly flat space-time. In this case, the static part and perturbation part can be separated. Metric tensor(Metric tensor is used in coordinate transformation) then can be written as:

$$g_{\mu\nu} = \eta_{\mu\nu} + h_{\mu\nu} \tag{3.1}$$

Where $\eta_{\mu\nu}$ is the Minkowski metric (Minkowski metric is a metric tensor of 4-dimentional spacetime, whose spatial coordinate is Cartesian.) diag(-1,1,1,1), representing a static part, and $h_{\mu\nu}$ is a small perturbation of smooth background simulating waves. The amplitude of waves is small, so $h_{\mu\nu} \ll 1$.

In short-wave approximation, the background metric $h_{\mu\nu}$ has a larger curvature radius than the wavelength. Moreover, the metric is time-dependent and changes slowly. The most useful approximation is the post-Newtonian approximation, where waves arise when corrected in high order. The correction sets general relativity away from its Newtonian limit [10].

3.2. Transverse-Traceless (TT) gauge and GWs' Polarization

In linearized theory, a specific gauge is chosen to make metric perturbation purely spatial, namely

$$h_{tt} = h_{ti} = 0 \tag{3.2}$$

Transverse-traceless meter can be introduced:

$$\partial_i h_{ij} = 0 \tag{3.3}$$

A metric perturbation can be written as $\mathrm{h}_{\mathrm{ab}}^{\mathrm{TT}}$ when satisfying TT gauge. "i" and "j" here are related to space coordinates.

In vacuum, the linearized Einstein field equation is

$$h_{ij}^{TT} = 0 \tag{3.4}$$

A detailed derivation of (3.4) is shown in reference [11]. Assume a GW traveling along z-axis. Perturbation depending on z and time t can be a solution for (3.4). In this case, the metric perturbation can be $\mathrm{h}_{\mathrm{ij}}^{\mathrm{TT}}(\mathrm{t}-\mathrm{z})$. From TT gauge, we have

$$\partial_z h_{zj}^{TT} = 0 \tag{3.5}$$

(3.5) tells that $\mathrm{h}_{\mathrm{zj}}^{\mathrm{TT}}$ is irrelevant to z. "t" and "z" are associated in $\mathrm{h}_{\mathrm{ij}}^{\mathrm{TT}}(\mathrm{t}-\mathrm{z})$, so $\mathrm{h}_{\mathrm{zj}}^{\mathrm{TT}}(\mathrm{t}-\mathrm{z})$ is also unrelated to t and must be a constant. To meet the condition that $\mathrm{h}_{\mathrm{ij}} \longrightarrow 0$ as $\mathrm{r} \rightarrow \infty$, $\mathrm{h}_{\mathrm{zj}}^{\mathrm{TT}}(\mathrm{t}-\mathrm{z}) = 0$ is accepted. Given this and (3.2), the non-zero components of h^{TT} are $\mathrm{h}_{\mathrm{xx}}^{\mathrm{TT}}$,$\mathrm{h}_{\mathrm{xy}}^{\mathrm{TT}}$,$\mathrm{h}_{\mathrm{yx}}^{\mathrm{TT}}$ and $\mathrm{h}_{\mathrm{yy}}^{\mathrm{TT}}$. From symmetry, we have

$$h_{xx}^{TT} = -h_{yy}^{TT} \equiv h_+(t-z) \tag{3.6}$$

$$h_{xy}^{TT} = h_{yx}^{TT} \equiv h_\times(t-z) \tag{3.7}$$

From TT gauge and Einstein field equation, we can get (3.6) and (3.7), representing the two modes of GWs polarization.

3.3. Perturbation and Strain

Consider two particles separated on x-axis. One is located at the origin, and another is located at $x = L_c$. The line element is

$$\mathrm{d}l^2 = g_{ij} dx^i dx^j \tag{3.8}$$

The distance between the two particles can be written as

$$L = \int_0^{Lc} \sqrt{g_{xx}(dx)^2} = \int_0^{Lc} \sqrt{g_{xx}}dx \tag{3.9}$$

When a GW propagates along z-axis, perturbation happens, and metric tensor g_{ij} has the form (3.1). (3.9) becomes

$$L = \int_0^{Lc} \sqrt{1 + h_{xx}^{TT}(t-z)dx} \approx \int_0^{Lc} \left[1 + \frac{1}{2}h_+(t-z)\right] dx = L_c\left[1 + \frac{1}{2}h_+(t-z)\right] \tag{3.10}$$

Because of the perturbation, particle distance changes in one direction of GW's polarization. The proportion of change is

$$\frac{\delta L_x}{L} \approx \frac{1}{2}h_+(t-z) \tag{3.11}$$

The same is true for the situation that two particles are separated on y-axis and a gravitational wave travels along z-axis:

$$\frac{\delta L_y}{L} \approx -\frac{1}{2}h_+(t-z) \tag{3.12}$$

Strain of GW can be defined as

$$h = \frac{\Delta L}{L} \tag{3.13}$$

And it is associated with the amplitude of the gravitational wave. (3.13) and (3.11) have similar form. It is natural to think that perturbation h_{xx}^{TT} works as a strain [11].

4. Indirect Detection–PSR 1913+16

The first evidence of gravitational waves is detected in the binary pulsar PSR 1913+16. In this binary system, there are 2 neutron stars, one of which is a pulsar. Pulsar makes the system perfect for testing general relativity. By fitting arrival times of pulse, component masses and size of orbit can be calculated. Combining them with orbital period P_b and eccentricity e, we can get the orbital period decay $\dot{P}_b$ rate. J. H. Taylor and J. M. Weisberg measured that $\dot{P}_b = (-2.30 \pm 0.22) \times 10^{-12}$[12] (more detailed calculations are contained in this reference). This measurement is consistent with the prediction of gravitational waves from general relativity.

Today, a program named TEMPO can analyze TOAs (time of arrival) data and output orbital parameters: projected semimajor axis $a_p sini$, e, epoch of periastron passage T_0, P_b, and longitude of periastron ω_0. It can also output post-Keplerian parameters, including orbital period derivative $\dot{P}_b$ [13].

5. GW Detectors

5.1. Resonant Detector

A resonant detector consists of a bar, suspensions, a resonant transducer, and an amplifier. Due to polarization, gravitational wave forces the bar to change in length and oscillate. Assume a bar with length $L\sim 1m$ and mass $M = 1000kg$. Using (3.13), we can calculate the amplitude of bar's vibration caused by a short burst gravitational wave with h $\sim 10^{-21}$:

$$\delta l_{gw} \sim hl \sim 10^{-21}m \tag{5.1}$$

where h is the strain of GW.

Resonant detector experiences three noises: thermal noise, sensor noise, and quantum limit.

Thermal noise is the most serious source of the noise. Place the bar in the environment with temperature T. When the kinetic energy of the bar equals to the thermal energy of a single degree of freedom, amplitude caused by thermal noise can be derived

$$\frac{M(\delta i)^2}{2}=\frac{kT}{2}$$

$$\langle\delta l^2\rangle_{th}^{\frac{1}{2}}=\left(\frac{kT}{4\pi^2Mf^2}\right)^{\frac{1}{2}} \tag{5.2}$$

If $T=100mK$, then $\delta l_{th}\sim 6.10^{-18}m$. δl_{th} is much larger than δl_{gw}. GW cannot be measured accurately under large noise. To detect GW, we need to increase Q, defined as $Q=f\cdot\tau$. f is the resonant frequency, and τ is vibrations' decay time. Due to random walks, the amplitude is $Q^{\frac{1}{2}}$ times smaller than the expected amplitude from (5.2):

$$\langle\delta l^2\rangle_{th}^{\frac{1}{2}}=\left(\frac{kT}{4\pi^2Mf^2Q}\right)^{\frac{1}{2}}\sim 6\cdot 10^{-21}m \tag{5.3}$$

It has the same order as the GW's amplitude. Bars nowadays can be operated at $\delta l=\delta l_{gw}+\delta l_{th}\sim 10^{-20}m$ [10].

The resonant transducer transforms vibration into an electric signal, and the amplifier enlarges the call to a recorded level. The amplifier then introduces **sensor noise**. The amplitude of vibration is the largest near the resonant frequency so that sensor noise can be ignored. Sensor noise limits the detector's sensitivity to the resonant frequency.

Quantum limit is caused due to uncertainty principle. It causes the vibration with an amplitude $\langle\delta l^2\rangle_{quantum}^{\frac{1}{2}}\sim 4.10^{-21}m$. Quantum limit is dominant when thermal noise is solved.

Bars are designed to detect bursts and can observe frequencies higher than 500Hz. Usually, the resonant detector is a cylinder, which is only sensitive to one direction along the principal axis. There is also a spherical detector, which is sensitive to all orders and can observe all sky.

5.2. *Interferometer*

Interferometers are based on interference, the same principle as the Michelson-Morley Experiment. There are two types. One is ground-based, and another is space-based.

5.2.1. Ground-Based Observatories. LIGO and Virgo are ground-based observatories. They detect gravitational waves in the frequency range between a few Hz and a few kHz [14,15]. When the laser meets the beam splitter in the center of the detectors, it is divided into two perpendicular beams of light. The original lengths of the two light paths are the same. When the gravitational waves come, the size of one way will be changed. Two beams of light have different phases when they meet again. At the photodetector, interference happens.

A GW with $h\sim 10^{-21}$ will change the length of one arm in interferometer

$$\delta l_{gw}\sim hl \tag{5.4}$$

Set $l=4m$, then $\delta l_{gw}\sim 4.10^{-18}m$. The typical period of gravitational waves of interest is much more extended than when light travels one round in arm. Hence, keeping light in arms and making light run 100 games is possible. Length change becomes $\delta l_{gw}\sim 4.10^{-16}m$, which makes GW easier to detect. Reflection is a way to keep light in your arms. Fabry-Pérot cavities, optical cavities with low-transmissivity mirrors, can achieve it.

Ground-based interferometer experiences five types of noise: ground vibration, thermal noise, shot noise, quantum effect, and gravity gradient noise [16].

Vibration caused by ground motion is much larger than that caused by a GW. To avoid this noise, one can set mirrors on pendulums, which make ground noise small at high frequency and offer isolation.

The pendulum introduces **thermal noise** at a few Hz. Light bounces between mirrors, causing internal vibrations of mirrors with natural frequencies at a few kHz. These can be limited to a narrow frequency band. Hence they can have a small impact on GWs measurement. Besides, some mirrors are partly transmissive. During transmission, mirrors absorb light's energy and increase temperature. As a result, the index of refraction changes, affecting interference. This effect limits the laser power used in detection.

Photons are quantized, so they arrive at the photodetector randomly and make random changes in light intensity, similar to that caused by a GW. This mechanism causes deviation (**shot noise**) in length measurement.

$$\delta l_{shot} \sim \frac{\lambda}{2\pi\sqrt{N}} \tag{5.5}$$

where N is the number of photons and λ is the light wavelength. The detected frequency is f, then the frequency of measurement is $2f$.

$$\mathrm{N} = \frac{total\ energy}{single\ photon\ energy} = \frac{P.\frac{1}{2f}}{\frac{hc}{\lambda}} \tag{5.6}$$

P is light power. Associating (5.5) and (5.6), a suitable method to reduce shot noise can be obtained: raising laser power. By setting a power-recycling mirror in front of the laser, photons' utilization can be increased, which has the same effect as improving strength.

According to the uncertainty principle, when length measurement is improved (shot noise is reduced), uncertainty in momentum measurement increases. **Quantum effects** are introduced. However, we don't need to know everything in the system. We only need to improve the measurements of quantities associated with GWs.

When GWs come, the gravitational field changes, leading to some natural sources like seismic waves and oscillations in air pressure. Detectors will respond to these sources and introduce **gravity gradient noise**. The noise has a huge impact when the frequency is low, so ground-based sensors are unsuitable for detecting frequencies below 1Hz.

5.2.2. Space-Based Observatories. One can launch the interferometer into space to avoid gravity gradient noise and detect GWs with a frequency below 1Hz [16].

LISA, TianQin, Taiji, and DECIGO are space-based observatories along Earth's orbit. (Diagrams of them are shown in references [17-20]. They share the same principle as ground-based observatories. In LISA, TianQin, Taiji, and spacecraft form triangles. DECIGO has four clusters, and each group has three satellites. The lengths of triangles are much longer than paths on the ground, so δl is more significant (according to (3.13)), which makes low-strain gravitational waves easier to observe. The minimum frequency that space-based observatories can detect is 0.1mHz.

5.3. Pulsar Timing Arrays

The arrival times of a pulse can be predicted by combining the pulsar's astrometric and orbital parameters, pulse propagation conditions, and spin. Comparing the prediction with observed TOAs, we can get the pulsar 'timing residuals', which represent space-time disturbances and is introduced by gravitational waves [21].

Pulsar timing arrays (PTAs) is a well-timed millisecond pulsars system that is perfect for detecting TOAs. The accuracy of its observation can reach 30ns. Therefore, PTAs can be used to measure timing residuals and, further, see gravitational waves. In order to improve the sensitivity of PTAs, (1) continuous observation of pulsars is required; (2) white and red noise needs to be reduced; (3) more pulsars need to be added.

Pulsar timing arrays are expected to detect ultra-low frequency($\sim 10^{-9} - 10^{-8} Hz$) GWs. The primary sources of it are massive black hole binaries in stable orbits with periods of few or tens of years, stochastic background gravitational radiation, and GW bursts [22].

6. CMB Polarization

CMB (cosmic microwave background), left after the Big Bang, has 2 modes of polarizations: E mode and B mode, caused by scalar and tensor perturbations (gravitational waves) in the early universe separately. Hence, observing B-mode polarization is a possible way to detect primordial gravitational waves.

Three devices can be used in B-mode polarization detection: ground-based observatories, high-altitude balloons, and satellites. Ground-based observatories ([23] gives an example of ground-based observatories.) are allowed to equip large-diameter telescopes, which lead to high resolution. Balloon-borne experiments avoid interfering emissions from the atmosphere and can be improved quickly. Like balloons, satellites are also free from atmospheric emissions. They can detect large angular-scale fluctuations but have limited resolution due to small telescopes.

To make observation more convincing, three issues need to be handled [24]. The first is improving sensitivity. Detectors are suffering thermal emissions from telescopes and the environment. Only satellites in space are cold enough to prevent this emission. Fundamental photon noise is the prominent noise of the best detectors. Therefore, the sensitivity of a single detector cannot be improved much. One can increase instrument complexity and the number of sensors to get higher sensitivity. The second is to identify CMB from diffuse emission from Milky Way. One can do the measurement at multiple frequencies. Then, the emission can be removed with the knowledge that galactic emission has a different spectrum than CMB. The third is to control instrument systematic uncertainties. With the improvement of equipment, a systematic errors can be reduced.

7. Conclusion

One way to generate gravitational waves is through the motions of massive bodies. It can produce GWs with frequencies from $10^{-4} Hz$ to $10^{4} Hz$. One joint motion is a coalescence of compact binaries containing 3 phases: inspiral, merger, and ring-down. What's more, due to instabilities and asymmetry, stellar core collapse can lead to the emission of GWs. A mismatch of magnetic axis and angular momentum makes neutron stars a source for continuous GWs. Gravitational waves can also be produced around inflation in the early universe due to fluctuations and inhomogeneities. These primordial GWs have an extensive frequency range.

Three assumptions are made in general relativity to get a mathematical description of GWs: linearized theory, short-wave approximation, and post-Newtonian theory. Introducing small perturbation $h_{\mu\nu}$ in spacetime and combining approximations with TT gauge, 2 modes of polarizations, "plus" and "cross", can be derived. Perturbations express polarization and act like strains.

A signal from PSR 1913+16 first proved the existence of gravitational waves. By fitting the data of pulse arrival time, we got binary's orbital period decay, which showed that the system emitted energy in the forms of GWs. The decay rate is consistent with general relativity. Today, TEMPO is designed to analyze TOAs.

The resonant detector is sensitive in a narrow band around its resonant frequency. It suffers thermal noise, sensor noise, and quantum limit. Sources with frequencies higher than 500Hz are suitable to detect using the deep detector. Interferometer is based on interference. Ground-based observatory detects GWs between a few Hz and a few kHz. It experiences ground vibration, thermal noise, shot noise, quantum effect, and gravity gradient noise. The last one limits its detectable frequency. To solve this problem, one can launch the observatory into space. Space-Based observatories can detect frequency below 1Hz; its minimum appreciable frequency is 0.1mHz. Pulsar timing arrays consist of well-timed millisecond pulsars. By analyzing their timing residuals, one can get information on GWs. PTAs can detect ultra-low frequency GWs.

Primordial GWs produced B-mode CMB polarization. Ground-based observatories, high-altitude balloons, and satellites detect B-mode polarization. Three issues need to be settled: (1) sensitivity, (2) identification of CMB, (3) systematic uncertainties.

With equipment and analysis methods improved, more detailed information in GWs signal can be explored. Through GWs, We look forward to discovering new theories corresponding to black hole properties, the early universe, and so on.

References

[1] Y. Gong and S. Hou, "The Polarizations of Gravitational Waves," Universe, vol. 4, no. 8, p. 85, Aug. 2018, doi: 10.3390/universe4080085.

[2] C. Bambi, "Astrophysical Black Holes: A Compact Pedagogical Review," ANNALEN DER PHYSIK, vol. 530, no. 6, p. 1700430, Jun. 2018, doi: 10.1002/andp.201700430.

[3] B. P. Abbott et al., "Observation of Gravitational Waves from a Binary Black Hole Merger," Phys. Rev. Lett., vol. 116, no. 6, p. 061102, Feb. 2016, doi: 10.1103/PhysRevLett.116.061102.

[4] E´. E´. Flanagan and S. A. Hughes, "The basics of gravitational wave theory," New J. Phys., vol. 7, pp. 204–204, Sep. 2005, doi: 10.1088/1367-2630/7/1/204.

[5] C. Cutler, "Gravitational waves from neutron stars with large toroidal B fields," Phys. Rev. D, vol. 66, no. 8, p. 084025, Oct. 2002, doi: 10.1103/PhysRevD.66.084025.

[6] J.-F. Dufaux, A. Bergman, G. Felder, L. Kofman, and J.-P. Uzan, "Theory and numerics of gravitational waves from preheating after inflation," Phys. Rev. D, vol. 76, no. 12, p. 123517, Dec. 2007, doi: 10.1103/PhysRevD.76.123517.

[7] R. Easther and E. A. Lim, "Stochastic gravitational wave production after inflation," J. Cosmol. Astropart. Phys., vol. 2006, no. 04, pp. 010–010, Apr. 2006, doi: 10.1088/1475-7516/2006/04/010.

[8] R. Easther, J. T. Giblin, and E. A. Lim, "Gravitational Wave Production at the End of Inflation," Phys. Rev. Lett., vol. 99, no. 22, p. 221301, Nov. 2007, doi: 10.1103/Phys- RevLett.99.221301.

[9] C. Bambi, "Observational Facilities," in Black Holes: A Laboratory for Testing Strong Gravity, Singapore: Springer Singapore, 2017, pp. 89–110. doi: 10.1007/978-981-10-4524-0 5.

[10] B. F. Schutz and F. Ricci, "Gravitational Waves, Sources, and Detectors." arXiv, May 26, 2010. Accessed: Mar. 04, 2023. [Online]. Available: http://arxiv.org/abs/1005.4735

[11] E´. E´. Flanagan and S. A. Hughes, "The basics of gravitational wave theory," New J. Phys., vol. 7, pp. 204–204, Sep. 2005, doi: 10.1088/1367-2630/7/1/204.

[12] J. H. Taylor and J. M. Weisberg, "A new test of general relativity - Gravitational radiation and the binary pulsar PSR 1913+16," ApJ, vol. 253, p. 908, Feb. 1982, doi: 10.1086/159690.

[13] J. M. Weisberg, D. J. Nice, and J. H. Taylor, "Timing Measurements of the Relativistic Binary Pulsar PSR B1913+16," ApJ, vol. 722, no. 2, pp. 1030–1034, Oct. 2010, doi: 10.1088/0004-637X/722/2/1030.

[14] B. P. Abbott et al., "LIGO: the Laser Interferometer Gravitational-Wave Observatory," Rep. Prog. Phys., vol. 72, no. 7, p. 076901, Jul. 2009, doi: 10.1088/0034-4885/72/7/076901.

[15] F. Acernese et al., "Status of Virgo," Class. Quantum Grav., vol. 22, no. 18, pp. S869–S880, Sep. 2005, doi: 10.1088/0264-9381/22/18/S01.

[16] B. S. Sathyaprakash and B. F. Schutz, "Physics, Astrophysics and Cosmology with Gravitational Waves," Living Rev. Relativ., vol. 12, no. 1, p. 2, Dec. 2009, doi: 10.12942/lrr-2009-2.

[17] K. Danzmann and A. R diger, "LISA technology concept, status, prospects," Class. Quantum Grav., vol. 20, no. 10, pp. S1–S9, May 2003, doi: 10.1088/0264-9381/20/10/301.

[18] J. Luo et al., "TianQin: a space-borne gravitational wave detector," Class. Quantum Grav., vol. 33, no. 3, p. 035010, Feb. 2016, doi: 10.1088/0264-9381/33/3/035010.

[19] Z. Luo, Z. Guo, G. Jin, Y. Wu, and W. Hu, "A brief analysis to Taiji: Science and technology," Results in Physics, vol. 16, p. 102918, Mar. 2020, doi: 10.1016/j.rinp.2019.102918.

[20] S. Sato et al., "The status of DECIGO," J. Phys.: Conf. Ser., vol. 840, p. 012010, May 2017, doi: 10.1088/1742-6596/840/1/012010.
[21] G. Hobbs et al., "The International Pulsar Timing Array project: using pulsars as a gravitational wave detector," Class. Quantum Grav., vol. 27, no. 8, p. 084013, Apr. 2010, doi: 10.1088/0264-9381/27/8/084013.
[22] A. N. Lommen, "Pulsar timing arrays: the promise of gravitational wave detection," Rep. Prog. Phys., vol. 78, no. 12, p. 124901, Dec. 2015, doi: 10.1088/0034-4885/78/12/124901.
[23] H. Li et al., "Probing primordial gravitational waves: Ali CMB Polarization Telescope," National Science Review, vol. 6, no. 1, pp. 145–154, Jan. 2019, doi: 10.1093/nsr/nwy019.
[24] L. M. Krauss, S. Dodelson, and S. Meyer, "Primordial Gravitational Waves and Cosmology," Science, vol. 328, no. 5981, pp. 989–992, May 2010, doi: 10.1126/science.1179541.

Rainbow link beamforming technology and relevant analysis

Yuheng Ding[1, 3, 4], Xueqing Zhang[2, 5]

[1]School of Communications and Information Engineering, Nanjing University of Posts and Telecommunications, Nanjing, 210003, China
[2]School of Electronic Information and Electrical Engineering, Shanghai Jiao Tong University, Shanghai, 200240, China
[3]Corresponding author

[4]1531606284@qq.com
[5]2416393284@qq.com

Abstract: In the 5G era, millimeter wave has been mentioned as an important position. Millimeter wave has some problems that cannot be ignored: the free space path loss and atmospheric attenuation. To overcome these problems, millimeter wave systems rely on beamforming to overcome losses. Traditional techniques utilize phase antenna arrays for beamforming, which can cause delays in the time consumption and beam scheduling, and due to limited beams, this technology can only be connected to a limited number of devices, and each transmission occupies the entire bandwidth. This paper introduces a new technology called rainbow link, which is implemented by a new true time delay (TTD) array structure. This technology can achieve the allocation of different subcarriers to a large number of users in the corresponding beamforming direction. Users only need to select appropriate frequency resources. In this work, we calculate the performance of rainbow link with different parameters, and our results show that given 1 GHz bandwidth using 64 element antenna arrays, a rainbow-link cell achieves 1ms latency and 0.6M symbols per second with 0.03 active user rate.

Keywords: Millimeter Wave Networks, Beamforming Technology, Massive Connection, Low Latency, True-Time-Delay Array.

1. Introduction

Millimeter wave (mmWave) strictly refers to electromagnetic waves with a wavelength between 1 and 10 millimeters and a frequency range of 30 GHz to 300 GHz [1]. Due to the lack of actual market demand for millimeter wave in the past, as well as the problems of large transmission loss, small coverage, and high component cost, millimeter wave has not been widely used [2]. However, in the 5G era, millimeter wave has been mentioned as an important position. After investigating the reasons, we believe that there are mainly the following points:

(1) Insufficient spectrum resources in the mid and low frequency bands
(2) 5G's demand for large bandwidth and high speed
(3) Good directionality and high transmission efficiency of millimeter wave transmission
(4) Millimeter wave communication has strong security and reliability, and high transmission quality

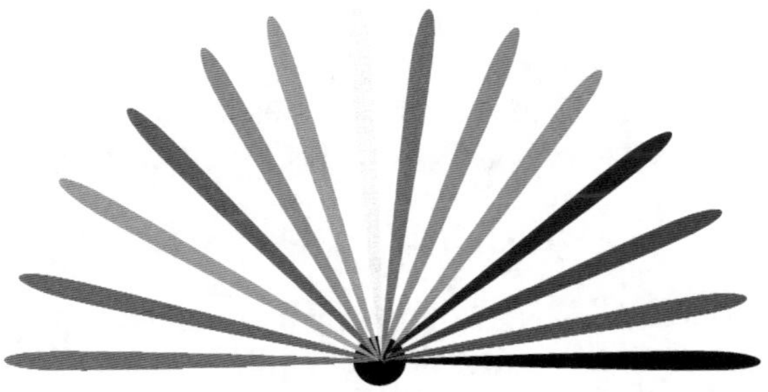

Figure 1. The beam formed using rainbow link technology.

Millimeter wave has some problems that cannot be ignored [3]: 1. The free space path loss is many orders of magnitude higher than the attenuation of traditional communication frequencies; 2. Atmospheric attenuation is large. To overcome these problems, millimeter wave systems rely on beamforming to overcome losses.

Beamforming is a concept derived from adaptive antennas [4]. The desired ideal signal can be formed by the signal processing at the receiverer by weighting and synthesizing the signals received by multiple antenna array units. From the viewpoint of antenna pattern, this is equal to forming a beam with a prescribed direction. For example, the original omnidirectional receiving pattern is converted into a lobe pattern with zero points and maximum direction. The transmitter applies the same principle. Adjust the amplitude and phase of the antenna element feed to form the desired shape of the pattern.

However, traditional beamforming techniques have significant limitations. Traditional techniques utilize phase antenna arrays for beamforming, and the time consumption associated with beamforming and beam scheduling can cause delays [5]. In addition, due to limitations such as the number of beams, this technology can only be connected to a limited number of devices, and each beamforming transmission occupies the entire bandwidth. Therefore, traditional beamforming does not support FDM. There are a large number of unlicensed frequency band resources available in the millimeter wave band [2], but due to limitations of traditional technologies, they cannot be shared by a great number of users.

From the existing research results, there are roughly two solutions to address the shortcomings of traditional technologies: optimizing the framework structure and all digital arrays. Although full digital arrays can achieve multi user connectivity, their power consumption and cost are very high. Therefore, a new technology called rainbow link [6] has been proposed, which is implemented by a new true time delay (TTD) array structure [5]. This technology can achieve the allocation of different subcarriers to a large number of users in the corresponding beamforming direction. Users only need to select appropriate frequency resources, thereby eliminating the need for beam training or scheduling. Therefore, unauthorized frequency band resources in the millimeter band are available and utilized. Next, our team will provide a detailed explanation of the principle and simulation of Rainbow Link.

2. Theory of Rainbow Link Technology

The novel rainbow link technology proposed by Ruifu L. and et al. [2] is a new method for beamforming technology, it applied the True Time Delay array [5] and is able to perform millimeter-wave communication, using an efficient orthogonal frequency-division multiplexing (OFDM) waveform with cyclic prefix (CP). Specific details will be briefly explained in this section.

We consider a massive input massive output (MIMO) situation where there're N_U antennas in the user's equipment (UEs) and N_B antennas in the base station (BS). The basics for rainbow link beamforming are adjusting the time delays and frequencies of waves in each antenna at the base station, and forming a wavefront pattern like in Figure 1. Each beam formed represented in different colours are in various frequencies and these sub-carriers are designed to be orthogonal, this increases the resource blocks available for the same bandwidth and the data capacity rises as well. To quantitively model the received signal of the UEs, we first define that for the *p*-th UE and *q*-th multipath component, the angle of arrival is θ_{pq} and the angle of departure is Φ_{pq}, these two variables are assumed to be distributed randomly in a normal pattern from π/2 to -π/2, the multipath gain is g_{pq}, the value is determined by the free space path loss model. Then we denote the complex gain for the *r*-th subcarrier as:

$$CG_{pqr}=g_{pq}e^{\frac{jr2\pi\tau_{pq}BW}{N_{SC}}} \tag{1}$$

Where *BW* stands for the total bandwidth of transmission and N_{SC} is the number of subcarriers, τ_{pq} is the multipath delay with reference to the first element of the antenna.

Meanwhile, we assume that the delays across different antenna elements in the same array are shorter than the duration of sampling, and thus can apply a narrowband response model. Therefore, the *k*-th narrowband array response vectors at the UE's side is given by $[ARU(\Phi)]_k = e^{j(k-1)\pi sin(\theta)}$, and that in the BS's side is $[\mathbf{AR_B}(\theta)]_k=e^{j(k-1)\pi sin(\Phi)}$. Hence, the channel **C** for a specific *p*-th UE and *r*-th subcarrier with N_M of multipath can be represented in the formula:

$$\mathbf{C}_{pr}=\frac{1}{\sqrt{N_B N_U}}\sum_{q=1}^{N_M} CG_{pqr}\mathbf{AR_U}(\Phi_{pq})\mathbf{AR_B^H}(\theta_{pq}) \tag{2}$$

Then, as each channel is at a rather small range of frequency, it's reasonable to assume it's flat over its bandwidth, and hence we may define the *t*-th element of the analog combiner of the beamforming array for the *r*-th subcarrier as $[\mathbf{w}_r]_t=e^{j(t-1)2\pi r/B}$. It is apparent that when the *r* changes, the angle of propaganda of the wave changes, this is why there may be various components of waves at different frequencies propagating at varied angles at the same time. And the term anchor sub-carrier is defined by the sub-carrier with the greatest beamforming gain at a certain angle.

Eventually, for a downlink transmission transmitting the OFDM signal $[T^{DL}]_r$ between the BS and the *p*-th UE at the *r*-th subcarrier, with the precoding array $\mathbf{P_p}$ at the UE side and additive white gaussian noise $[AWGN^{DL}]_r$ received signal $\mathbf{R_U}$ at the UE side is given by:

$$[\mathbf{R_U^{DL}}]_r=\mathbf{C_{pr}w_rP_p^H}[\mathbf{T^{DL}}]_r+[\mathbf{AWGN^{DL}}]_r \tag{3}$$

Similarly, the received signal $\mathbf{R_B}$ at the BS side can be formulated by:

$$[\mathbf{R_B^{UL}}]_r=\sum_{p=1}^{N_{AU}} \mathbf{C_{pr}^H w_r^H P_p}[\mathbf{T^{UL}}]_r+[\mathbf{AWGN^{UL}}]_r \tag{4}$$

Where N_{AU} stands for the number of active users per time frame.

3. Evaluation & introduction to the simulation

Here we have introduced a special transmission method, which is rainbow link technology. In order to verify its advantages and disadvantages, there are several evaluative points we can analyze. With the use of simulation, these factors can be presented completely with vivid graphs and data, hence we can make comparisons between other methods of beaming forming technology.

3.1. SNR

SNR is short for "Signal to Noise Ratio". It is the ratio of signal power to the power of all other electrical signals in the area, called the noise level. Noise is measured by the Root-Mean-Square (RMS) value of the fluctuations over time. This ratio is expressed in decibels (dB).

We consider SNR as a suitable factor for analyzing because SNR determines how usable the signal will be. A higher value of SNR means the signal is clearer. The noise interferes with the signal processing capabilities of the network, resulting in random noise and amplitude modulation. If the value of SNR gets lower than one, the signal becomes unusable. Due to different frequency and different pointing direction in the rainbow link technology, each subcarrier will have its own beam forming gain, which leads to its own SNR.

We have mentioned that SNR is expressed in the ratio of signal power to noise power. Here is how to calculate:

$$\mathrm{SNR} = \frac{\mathrm{Pr}}{\mathrm{Pn}} = \frac{\mathrm{Pt} * \frac{\mathrm{Gtx} * \mathrm{Gr}}{\mathrm{Lp}}}{\mathrm{N0} * \mathrm{b}} \tag{5}$$

where Pr is the receive power and Pn is the noise power. For further decomposition, receive power can be rewritten as the function above, where Pt, Gtx, Gr representing transmit power, gain of the antenna, and gain of the receiver respectively. Lp reveals the path loss during transmission. In terms of noise power, we denote the lower case *b* as the actual bandwidth that User Equipments(UE), or IoTs receive, instead of the total bandwidth that BS transmits.

In the last section, we have mentioned the detailed formula of calculating SNR. In our simulation which will be introduced in the later part, we decide to set transmit power (Pt) and N0 as constants. The determinants of the rest four parameters are as shown:

Table 1. Determinants of SNR.

Parameters	Determinants
Gain of the antenna(Gtx)	Number of antennas
Gain of the receiver(Gr)	Number of receivers
Path loss (Lp)	Distance
b	Number of SC, total bandwidth

As a result, it's possible for us to alter the other four variables in order to set different conditions for the simulation. It can help us to gain more accurate and comprehensive results through analyzing SNR, as well as making better conclusions in advantages and disadvantages of the rainbow link technology.

3.2. Contention and Packet Loss

Packet loss occurs when these packets don't reach their final destination – some of them can get lost in congested networks, which is known as "Packet Collision". In our daily lives, Packet Loss usually provides us with the information of the efficiency of the transmission of data between networks and devices. That's why we choose it as another suitable factor for evaluating in our simulation.

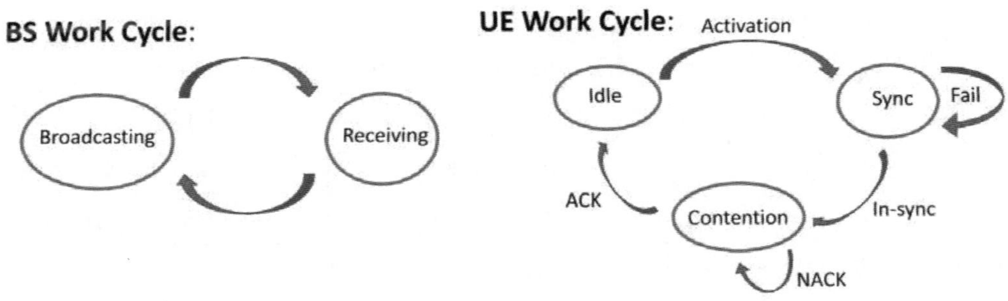

Figure 2. BS&UE Work Cycle [2].

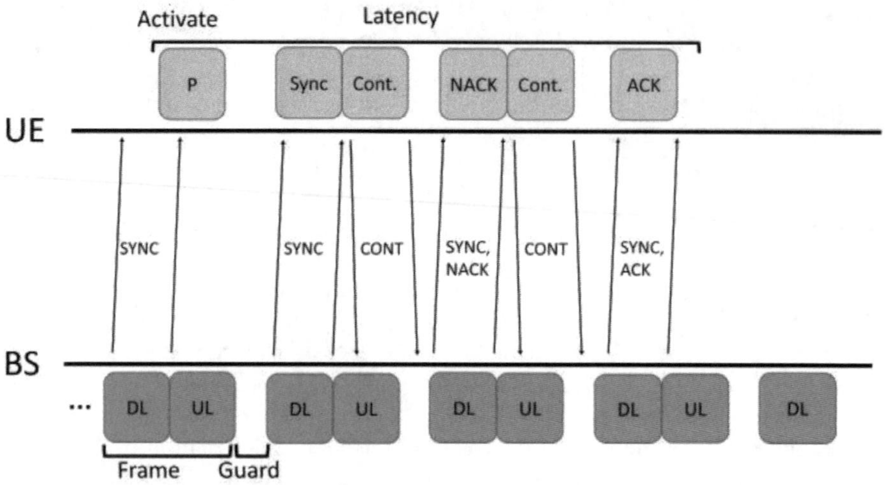

Figure 3. User Plane Latency [2].

In Figure 2, we can be informed that the base station act as transmitter first, and UE acts as receiver. BS will transmit synchronized packets of data as well as a time signal to UE. After that, BS will switch to become a listener and wait and see if any UEs, or IoTs want to send responsive data back. For UEs, they initially remain in the "Idle" mode, which means they are not working or sleeping. As soon as they are activated through any way, such as receiving the time signal or the synchronized packets of data, UEs will turn into the "Sync" mode. This mode simply reveals that UEs are kind of waiting for base station to transmit data, and they listen. While they fail to receive the synchronization signal, UEs will just go back and need to wait for the next time. Once they successfully find the synchronization signal, UEs will transmit. Since every IoT operates on its own, they may send data back to BS at the same time, and possibly the transmit data will fall in the same rainbow link. This will cause packet collision, or what we called contention to occur. While contention keeps remaining, negative acknowledged signal (NACK) will be kept sending by UEs. As soon as contention is solved, acknowledged signal (ACK) together with the synchronized data, will be sent back to BS. Until then UEs will switch back to "idle" mode and wait for the next work cycle.

Similar to SNR, there are also several factors that can influence the packet loss. We can derive these factors following the similar pattern. Here shows these factors:

Table 2. Determinants of Packet Loss / Contention.

Symbols of the determinants	Meanings
U	the number of users participating in the contention
K	the number of resource block in subcarriers.
B	the number of total resource blocks
n	the number of repetitions that each UE makes

In our later simulation part, we will focus on simulating and evaluating K and n.

4. Performance analysis and simulation result

In this section, we analyse the performance of rainbow link and support it with simulation results based on MATLAB. Table 3 lists the specific values of the system parameters we set in the simulation.

Table 3. Specification of system design.

Specification	Explanations	Value
BW	Total bandwidth of the network	1 GHz
NB	Number of antennas in BS	64
B	Total number of SC	2048
f_c	Centre frequency	60 GHz
P_t	Power of transmitter	20 dBm
G_{tx}	Antenna gain	18.1 dBm
N_0	PSD of noise	8*10-15 W/Hz
Frame length	Frame design	125 μs
UL payloads	Frame design	18 symbol

4.1. Gain Loss

In the previous section, it indicates that beamforming gain is frequency dependent due to TTD array, causing gain loss on different SCs. As shown in Figure 4, for a given segment of SCs, the edge SC (in case of yellow) has larger gain loss in comparison with the anchor SC (in case of blue) when IoT is from different directions. As a result, UEs needs to find the anchor SC with the highest gain to locate a segment of SCs with the minimum beamforming gain loss as its operating band *Bu*, which is defined by SC set a narrowband UE can access [2]. Owing to the gain loss, *Bu* can include only limited SCs. And through the comparison in Figure 4, it is clear that when number of antennas N_B at the BS increases, it might cause large gain difference as the beams will narrow with increased antennas N_B.

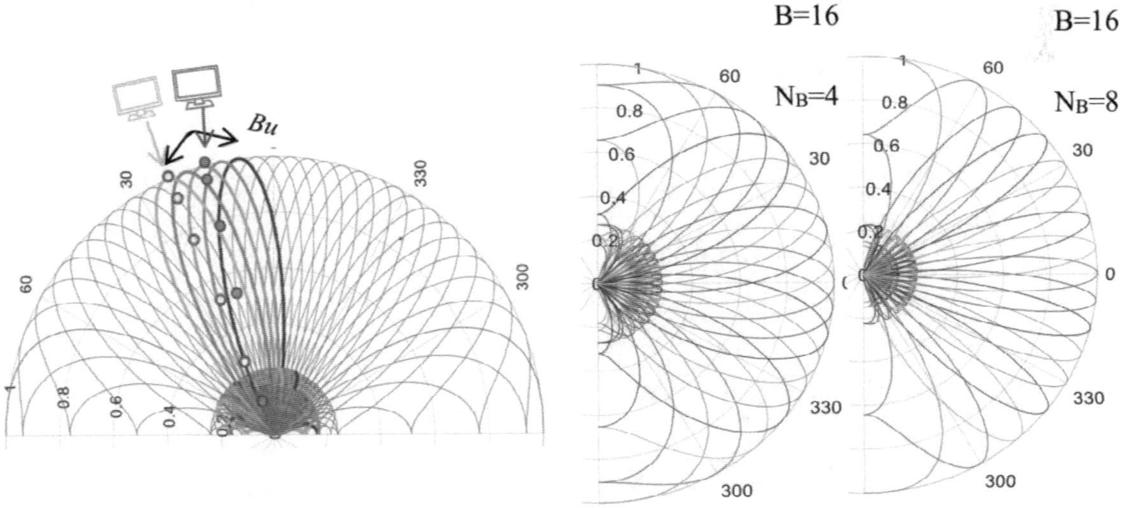

(a)Cause of gain loss.

(b)Beamforming patterns with different size of antenna.

Figure 4. Beamforming patterns.

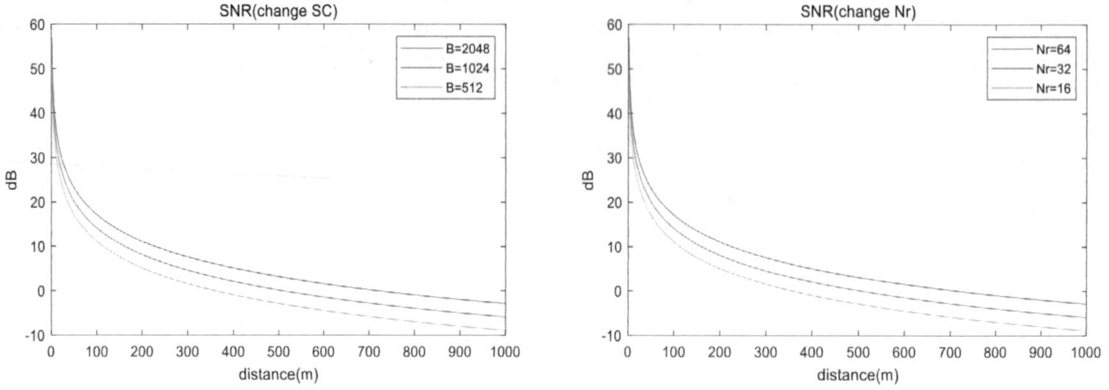

(a)SNR with different B and N_B.

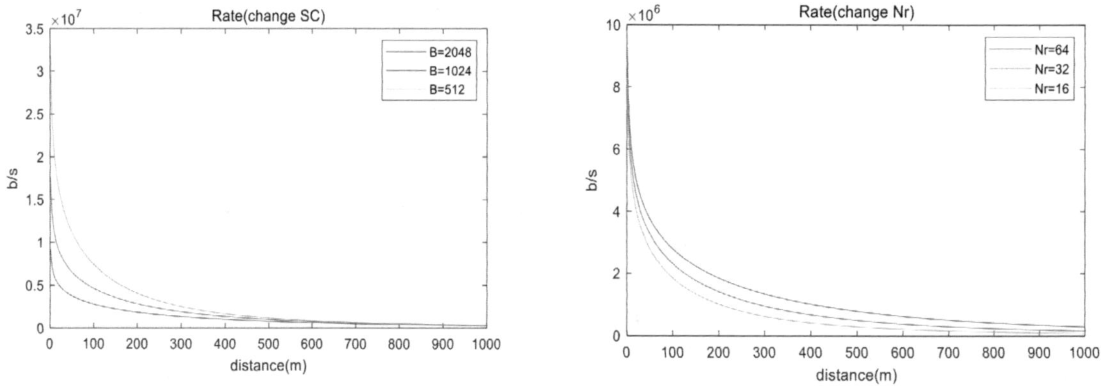

(b)Data rate with different B and N_B.

Figure 5. SNR and maximum transmission rate.

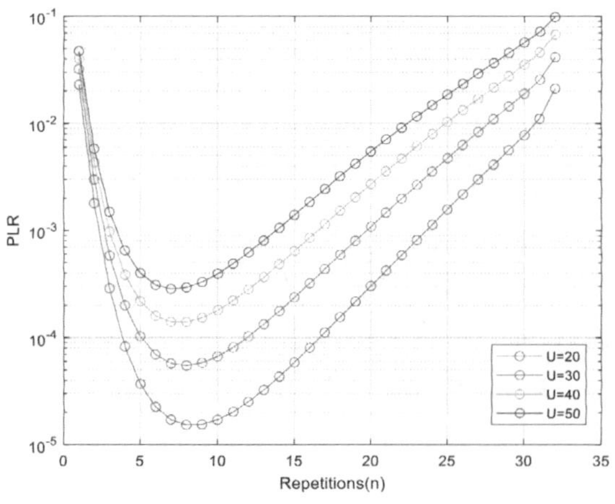

(a)Packet loss rate with $g_{group} = 1$.

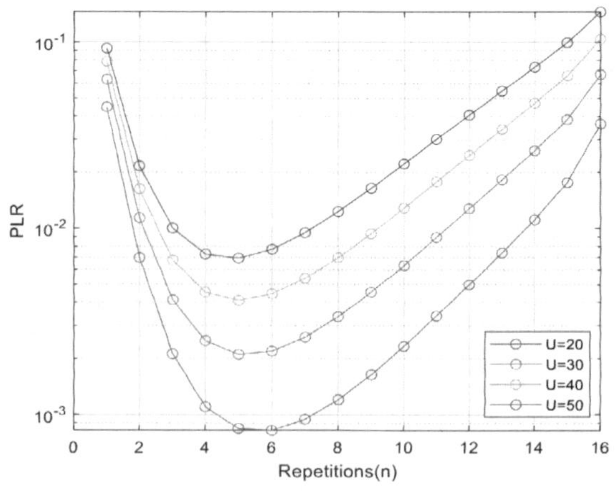

(b)Packet loss rate with $g_{group} = 2$.

Figure 6. Packet loss rate.

4.2. SNR and Data Rate

Figure 5 gives detailed comparison of a single UE's receiving SNR and maximum data transmission rate when changing the number of SCs B and antennas N_B separately. The SNR is given by Equation (5) and the data rate is given by Shannon formula. It shows that a UE will have higher SNR and lower rate with increased SCs B while it will have higher SNR and rate at the same time with increased antennas N_B. Specifically, users at 400m can only achieve 5dB SNR. As distances increases, path loss will become higher accordingly, resulting in higher SNR. So for users at marginal area of the cell, they have very low SNR which might affect synchronization performance as more interference exists.

4.3. Packet Loss Rate

Taking the formula in [2] as a reference, shown by Figure 6 we get the following result of packet loss rate(PLR) in case of g_{group} = 1,2, where g_{group} represents number of SCs bundled into a resource block(RB). Firstly, the result indicates that for given UEs, the P_{PLR} has a minimum value with a special repetitions n. And as active users U in contention increases, the optimal n decreases slowly. Secondly, the PLR with $g_{group} = 2$ is higher than that with $g_{group} = 1$. This phenomenon occurs because although a RB can carry more useful symbols in case of 2-grouping, there are less RBs to avoid collisions and thus increases P_{PLR}. Specifically, given active users $U = 20$, the PLR in 1-grouping can achieve about 10^{-5} with optimal $n = 8$ while in 1-grouping it can achieve about 10^{-3} with optimal $n = 6$.

4.4. Contention Latency and Effective Rate

Since the latency caused by synchronization is fixed, we mainly focus on the latency caused by UE's contention. The frame length is 125μs as listed in Table 1. Figure 7 shows that the line graph appears stepped due to multiple repetitions in contention. It illustrates that when more SCs are included in a RB, the latency increases for higher PLR while it promotes the effective rate of a single user. Specifically, the rate with 2-grouping is only half of that with 4-grouping in case of user active probability $p = 0.03$.

As shown in Figure 8, we change the design parameters with appropriate value to ensure the similar performance of latency, where $p = 0.18$ in case of $g_{group} = 1$ and $p = 0.03$ in case of $g_{group} = 4$ separately. From the previous result we can easily know that although with 1-grouping the effective rate of a single user achieves only one fourth of that with 4-grouping, it can serve 6 times as many users.

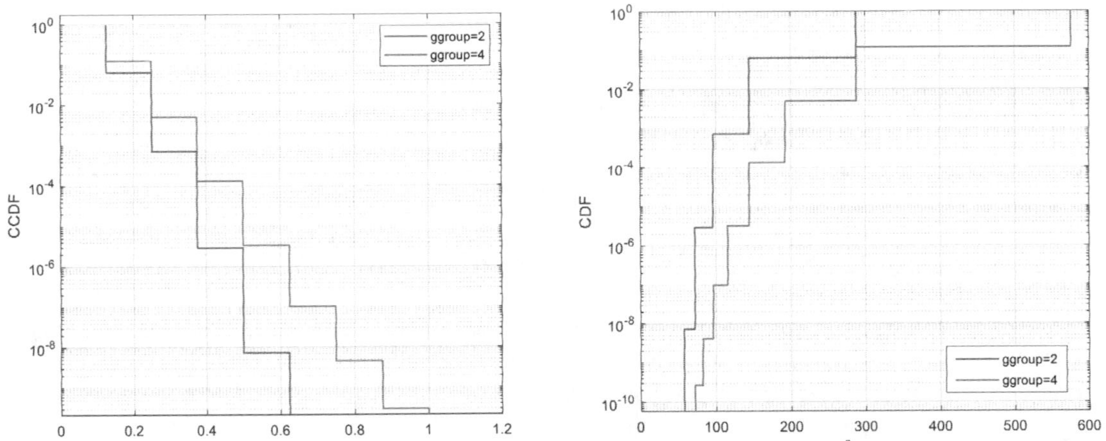

(a)Contention latency with $g_{group} = 2,4$. (b)Effective rate with $g_{group} = 2,4$.

Figure 7. Contention latency and corresponding effective rate with different g_{group}.

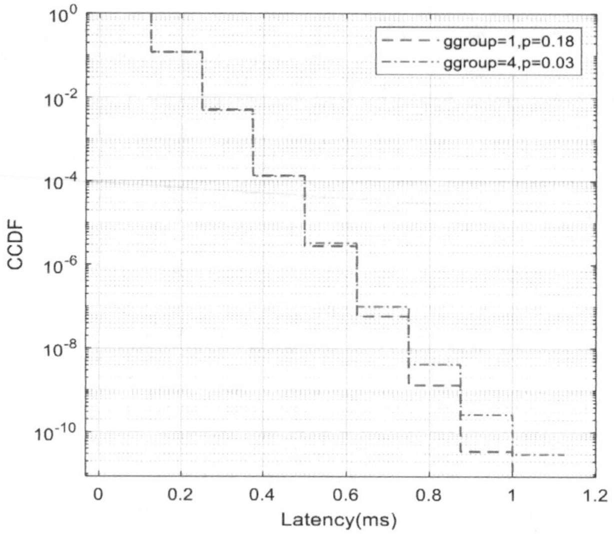

Figure 8. Contention latency with different parameters that ensure it similar.

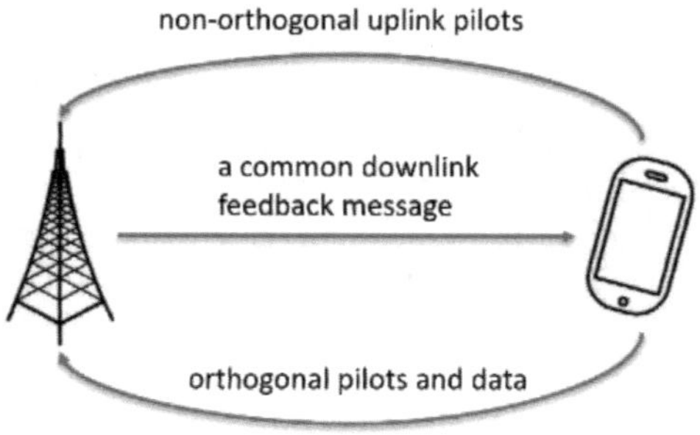

Figure 9. three steps of the scheduling-based scheme.

5. Introduction of another method to improve the multiple access

In this section, we briefly introduce another possible method to improve the multiple access of the rainbow link.

As discussed above, multiple packets transmitted in the same time-frequency resource block from different terminals would cause collision and packet loss. To reduce packet loss, we use a simple repetition coding. However, this method requiring abundant time-frequency resource blocks causes much waste. So maybe we can find a better approach.

The paper [7] proposed a scheduling-based method for massive random access in MIMO systems. The method is based on two discoveries that can significantly reduce downlink feedback for scheduling in order to avoid interference. Firstly, it suggests replacing the AMP (approximate message passing) with the covariance approach in cases where we only need to estimate the large-scale fading. This approach is more efficient in detecting active users and estimating their channels using fewer pilot symbols. Secondly, the feedback cost is relatively low, and coordination among active devices can be achieved through a common feedback message from the BS to the users. Moreover, the amount of feedback required for collision-free scheduling scales only linearly with K, with a small coefficient (K being the number of active users).

Therefore, it proposed a sourced approach - through an activity detection process, the BS can identify the active users, then it will schedule the active users by using a feedback strategy. In general, this method has three steps. Firstly, active users transmit non-orthogonal uplink pilots to BS for activity detection. After active users have been determined by BS, it broadcasts a common downlink feedback message to assign each user a slot and an orthogonal pilot. Finally, the users transmit their orthogonal pilots and data in the scheduled slot to the BS.

Here is a more detailed implementation.

Firstly, each of the N potential users are assigned uniquely identifying non-orthogonal pilot sequences $s_1, s_2, \ldots, s_n$. In the pilot phase, the active users transmit their pilots to the BS. The received signal can be expressed as:

$$Y^T = \sum_i s_i g_i h_i^T + Z \tag{6}$$

where $[s_1, \ldots, s_N]$ is the signature sequence matrix, g_i is the large-scale fading, h_i is the Rayleigh fading component, and Z is the AWGN noise.

At each frame end, the BS should use all Y_d, $d = 1, \ldots, \Delta$, to determine the active users among all potential users. There are two methods – AMP and the covariance approach. While the AMP algorithm is suited for moderate number of antennas, the covariance approach has a distinct advantage at great number of antennas, because it takes advantage of the channel hardening effect in the massive MIMO regime.

After identifying the set of active users, the BS could then send a collective feedback message to organize them within transmission slots in the frame. The most efficient feedback strategy involves creating a codebook of hashing functions. Each set of active users can then be mapped to distinct hashed values, by selecting a hash function from the list, and using the hash function index as a codeword. Once this is done, the users can transmit their orthogonal pilots, as well as their data, in the assigned slots.

The proposed method is analyzed in two scenarios - fast-fading and slow-fading - and compared to the contention-based method, with results in proving better performance. Significantly, the research emphasizes the merits of incorporating feedback in random access protocols, particularly given the relatively small cost. This implies that our rainbow link could see an improvement in performance with the application of this method.

6. Conclusion

In order to decrease delays in the time consumption and beam scheduling caused by traditional beamforming method, we presented a novel technique called rainbow link. The basis is to utilize the broad spectrum in millimeterwave and map different frequencies to specific directions, enabling coverage of the entire angular space. In this paper, we used MATLAB to analyse the performance of rainbow link, indicating that grant-free rainbow link access could viably support low-latency connectivity requirements. It can be seen from the analysis that as the number of subcarriers increases, there will be more beams, higher SNR and lower transmission rates. Besides, as the number of antennas in the Base Station increases, beams become narrower, and SNR and transmission rates will be higher. Moreover, one limitation of the rainbow link is the collision and packet loss in multiple packets transmission. Here we used a simple repetition coding to reduce this loss, which causes much waste. So, at the end of the article, we introduced a scheduling method for massive random access in MIMO systems. Maybe it could be applicable to improve the performance of rainbow link in the future.

References

[1] S. Rangan, T. S. Rappaport and E. Erkip, "Millimeter-Wave Cellular Wireless Networks: Potentials and Challenges," in Proceedings of the IEEE, vol. 102, no. 3, pp. 366-385, March 2014, doi: 10.1109/JPROC.2014.2299397.

[2] R. Li, H. Yan, and D. Cabric, "Rainbow-link: Beam-alignment-free and grant-free mmW multiple access using true-time-delay array," IEEE Journal on Selected Areas in Communications, vol. 40, no. 5, pp. 1692 – 1705, May 2022

[3] S. Chen and J. Zhao, "The requirements, challenges, and technologies for 5G of terrestrial mobile telecommunication," in IEEE Communications Magazine, vol. 52, no. 5, pp. 36-43, May 2014, doi: 10.1109/MCOM.2014.6815891.

[4] W. Roh et al., "Millimeter-wave beamforming as an enabling technology for 5G cellular communications: theoretical feasibility and prototype results," in IEEE Communications Magazine, vol. 52, no. 2, pp. 106-113, February 2014, doi: 10.1109/MCOM.2014.6736750.

[5] E. Ghaderi, A. S. Ramani, A. A. Rahimi, D. Heo, S. Shekhar, and S. Gupta, "An integrated discrete-time delay-compensating technique for large-array beamformers," IEEE Trans. Circuits Syst. I, Reg. Papers, vol. 66, no. 9, pp. 3296–3306, Sep. 2019

[6] H. Yan, V. Boljanovic, and D. Cabric, "Wideband millimeter-wave beam training with true-time-delay array architecture," in Proc. 53rd Asilomar Conf. Signals, Syst., Comput., Nov. 2019, pp. 1447–1452.

[7] J. Kang and Y. W. Yu, "Scheduling Versus Contention for Massive Random Access in Massive MIMO Systems," IEEE Transactions on Communications, vol.70, pp. 5811 – 5824, Sep. 2022

Application research in multi-objective constrained optimization of composite wing structure

He Yan[1,2], Cheng JiaLin[1], Mao Sen[1], and Zhang WeiJing[1]

[1]AVIC CHENG'DU AIRCRAFT INDUSTRIAL (GROUP) CO. LTD., Chengdu 610092, P. R. China

[2]171357852@qq.com

Abstract. In order to explore the optimization of composite wing structure, this paper reviews the past research works of optimization of composite wing structures worldwide and in China. It concludes that most optimization methods and calculations are useful but only applied with specified theoretic model. Then relevant composite optimization theories and strength constraints with composite lamination failure criteria, stability, aeroelasticity are sorts out. Optimizing algorithm and applied solution rounding strategy are also listed. Concentrating on weight-reduction analysis on the thickness, sequence, and angle of the composite stacking under multiple constraints, a practical model of a composite wing structure example with a large aspect ratio is completed, and the calculation is based on the finite element analysis tool Nastran and the genetic algorithm Matlab program. Two optimal solutions are given with detailed differences and the comparison of the composite structure parameters before and after calculation is also recorded. The application example indicates the feasibility of the optimization method and quadratic rounding process, the detail structure performance improvements with skin lamination change under the calculation are illustrated at the same time.

Keywords: Composite Stacking, Wing Box Structure, Genetic Algorithm, Multi-Objective Constrained Optimization.

1. Introduction

In recent years, the continuous progress of new material technology has promoted the development of aircraft design and manufacturing. The proportion of advanced composite materials in aircraft structures is also increasing. For example, composite materials in the Boeing 787 Dreamliner make up over 50% of the aircraft's weight. Composite materials have many advantages such as high specific strength, high specific modulus, good fatigue resistance, corrosion resistance, and superior designability. The proportion of advanced composite materials has become an important factor to measure the advancement of aircraft. In the aircraft structure, the wing is the most important structural part for the aircraft to generate lift. Wings made of composite materials can reduce the structural weight by 20% to 30% and improve fuel economy by 7% to 15%. The remarkable feature of composite materials is that they are anisotropic, and their deformation designability can be used to improve aeroelastic quality and the overall lift/drag characteristics of the aircraft on the basis of reducing structural weight. However, the extensive use of composite materials in structural design has led to a substantial increase in the number of relevant design parameters in the initial design. How to determine the optimal value of each parameter has become a major difficulty in wing design. Therefore, finding a set of design parameters that can minimize the structural mass while satisfying various performance indexes becomes an important task of wing structure design.

A lot of research has been carried out worldwide on the advantages of composite materials and the difficulties of applying them to structural design. In the late 1980s, Koji Isogai studied the aeroelastic tailoring of composite wings under multiple constraints by applying the direct search algorithm, avoiding the problems caused by discontinuous sensitivity information in sensitivity-based optimization [1]. S. Guo et al. used different optimization methods for aeroelastic tailoring of the composite wing structure, calculated the maximum flutter velocity of the wing after tailoring [2], and studied the effect of sweep angle, root-to-shoot ratio, and mass distribution on the aeroelastic tailoring of composite wings based on the gradient-based deterministic algorithm and genetic algorithm [3]. In 2012, J. K. S. Dillinger et al. used the in-plane and bending stiffness matrices and laminate thickness as design variables to optimize the stiffness of the upper and lower skins of a composite wing, and found the responses of the structure and their sensitivities to the design variables [4]. Bret K. Stanford et al. evaluated the effects of various aeroelastic tailoring schemes for subsonic transport wings by solving minimum-mass aeroelastic optimization problems [5]. S. Nagendra [6], J. Fatemi [7], and J. H. Kang [8] et al. applied the genetic algorithm to optimize the design of stiffened panels. W. Liu [9] et al. proposed an optimization method for composite stiffened panels using the program package VICONOPT.

In China, in 1989, Zou Congqing explained the basic principle of aeroelastic tailoring and its performance benefit [10]. Zhang Yining et al. proposed solutions for the static strength, equilibrium constraints, and dynamic limits in the optimization process, based on which they programmed to optimize a triangular composite wing. Tests on the optimized wing structure had consistent results with the optimization, achieving the intended goal [11]. Wan Zhiqiang et al. conducted an aeroelastic tailoring analysis on a forward-swept composite airplane using the genetic/sensitivity-based hybrid algorithm. The results indicated that applying the hybrid algorithm had better performance than using sensitivity-based optimization alone [12]. Cheng Wenyuan et al. used the Pareto multi-objective genetic algorithm to optimize the composite wing for the aileron reversal problem in an aircraft design, which greatly improved the optimization efficiency [13]. Lu Jie et al. optimized the structure of each part of the wing with Nastran by controlling the structural displacement, strain, and other conditions [14]. Liu Kelong [15] used low-degree-of-freedom collaborative optimization to optimize the multidisciplinary structural aeroelastic design of wings. Zhang Tieliang [16] realized the layout optimization of the stiffened panel by building a surrogate model in experimental design.

In the past few decades, a lot of research and exploration have been carried out on the optimization of composite wings at home and abroad. Compared with conventional design, targeted tailoring optimization of composite materials can obtain good weight reduction, which helps improve the efficiency of actual structural design, shorten the design time, and obtain optimal design parameters. However, there are various parameters and optimization methods for designing composite wing structures, each with its own advantages and disadvantages. At present, most of the optimization design still aims at specific goals and the optimization of composite wings is still one of the research hotspots for designers of aircraft structures. In this paper, the genetic algorithm is used. Considering multiple constraints such as static force and aeroelasticity, this paper optimizes the size and stacking of the composite wing structure on a certain type of aircraft, aiming at weight reduction.

2. Structural optimization and genetic algorithm

The optimization of composite wing structure design is a computational integration involving multiple subjects in mechanics, including classical lamination theory, stability, aeroelasticity, optimization algorithms, etc. Weight reduction is the most important design goal in optimization. The three optimization elements include design variables, design constraints, and design goals. According to different design variables, it can be divided into size optimization, shape optimization, and topology optimization. The optimal values of these variables are different but coupled with each other. Generally, the structural layout has the greatest impact on the weight of wings, but it is still immature comparing to size optimization.

As a composite material structure, the prepreg made of carbon fiber and resin is used as a mechanical material to be laid up as a laminate at different angles and thicknesses, which can be used as a typical basic load-bearing structure on the aircraft. Here, the structural optimization variables include stacking thickness, angle, and sequence. The stacking thickness is directly related to the weight of the laminate. The stacking angle is obtained by changing the angle to obtain the ideal stiffness, which is often used in aeroelastic tailoring design. Stacking sequence optimization keeps the in-plane stiffness of the laminate but changes the bending stiffness to improve its stability, vibration, and etc.

2.1. The mathematical model for composite wing structure optimization

Design variables include all independent characters of the structure that need to be determined in the design, including size, shape, topological parameters, etc. In a detail design of composite laminates, thickness, angles and stacking sequence are the main variables.

$$X = \{x_1 \dots x_n\}$$

The design target is the structure weight.

$$\min : F(X) = W(X)$$

And the muti-constraints must be satisfied.

$$G_i(X) = Y(X) - Y_{MAX} \le 0$$

Where $x_i^{min} \le x_i \le x_i^{max}$ is the boundary condition, when considering stacking angle, it is $\theta_i \in (0, 45, -45, 90)$. $W(X)$ is the objective function of weight and response constraint functions $G_i(X)$ are listed in Section 2.2 in detail while the margin Y_{MAX} should be satisfied.

2.2. Theories of strength constraints

a) lamination strength

Tsai-Wu Failure Criteria with more mechanic index is used in this paper when analyzing the composite laminate subjected to an external load in plane stress state.

$$F_{11}\sigma_1^2 + 2F_{12}\sigma_1\sigma_2 + F_{22}\sigma_2^2 + F_{66}\sigma_6^2 + F_1\sigma_1 + F_2\sigma_2 = 1$$

Where:

$$F_{11} = \frac{1}{X_t} - \frac{1}{X_c},\ F_2 = \frac{1}{Y_t} - \frac{1}{Y_c},\ F_{11} = \frac{1}{X_c X_t},\ F_{22} = \frac{1}{Y_c Y_t},\ F_{66} = \frac{1}{S^2},\ F_{12} = -\frac{\sqrt{F_{11}F_{22}}}{2}$$

X_t, X_c, Y_t, Y_c, S are the strength parameters of the each composite ply.

b) Stability of composite laminates

The buckling critical load for a typical rectangular orthotropic laminate with its four sides simply supported is calculated as:

$$N_x = \frac{\pi^2 D_{22}}{b^2}\left[\frac{D_{11}}{D_{22}}\left(\frac{b}{a}\right)^2 m^2 + 2\left(\frac{D_{12} + 2D_{66}}{D_{22}}\right) + \left(\frac{a}{b}\right)^2 \frac{1}{m^2}\right]$$

where m is the number of half-wavelength buckling in the unstable direction of the plate, and the minimum value in the load group of the calculation result is the buckling critical value of the plate; D_{11}, D_{12}, D_{22}, and D_{66} are the bending stiffness coefficients of the composite plate; N_x, N_y, and N_{xy} are the load acting on the unit length on the periphery of the plate.

c) Aeroelasticity
The aeroelastic flutter equation is:

$$[M]\ddot{q}+[K]q=[Q]$$

where M and K are the generalized mass and stiffness matrices of the wing structure, q is the wing displacement vector, and Q is the unsteady aerodynamic force.

In the critical state of wing flutter, the wing can be considered to be in approximate simple harmonic vibration. Using the p-k method, the flutter equation is rewritten as:

$$[(p^2[M]-p\frac{b}{2k}\rho V[Q]^I)+([K]-\frac{1}{2}\rho V^2[Q]^R)]q=0$$

where $p=\omega(\gamma+i)$ is the complex eigenvalue of the equation, γ is the attenuation rate; $[Q]^I$ is the aerodynamic damping matrix, $[Q]^R$ is the aerodynamic stiffness matrix; ρ is the air density, V is the incoming flow velocity, b is the characteristic length of the wing, and $k=\frac{b\omega}{V}$ is the shrinkage frequency, representing the interactions between motions at points.
Defining aileron manipulation efficiency:

$$\lambda=(\partial C_{mx}/\partial\delta_{\alpha})_{e}/(\partial C_{mx}/\partial\delta_{\alpha})_{r}$$

Where C_{mx} is the the rolling torque coefficient, δ_{α} is the angle of aileron deflection, "e" is the elasticity and "r" refers to rigidity. $\lambda\geq 1$ should be achieved when avoiding aileron reversal.

2.3. Genetic algorithm and rounding strategy
The genetic algorithm is an adaptive, global optimization, and probabilistic retrieval optimization algorithm formed by simulating the genetic and evolution process of organisms in the natural environment. The characteristics of the genetic algorithm mainly include coding, selective replication, crossover, and mutation. Its advantages mainly include using mathematical codes to represent decision variables and borrowing the concepts of chromosomes and genes in biology, which is beneficial for optimization problems that are difficult to express numerically. The random global search and optimization methods developed by imitating the biological evolution mechanism in nature draw on the theory of evolution and genetics. It automatically acquires and accumulates knowledge about the search space during the search process, which is adaptive controlled to obtain the optimal solution. The principle of "survival of the fittest" generates near-optimal methods successively among potential solutions.

This paper uses the genetic algorithm for the optimized calculation and analysis of the structure. The algorithm procedure is shown in figure 1.

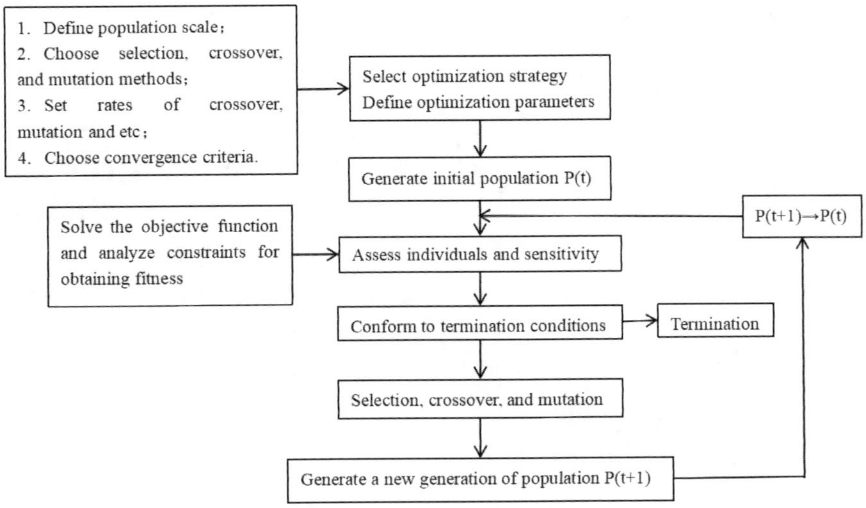

Figure 1. The genetic algorithm flowchart.

To obtain the optimal solution for composite wings with thinner stacking, quadratic optimization is easier. On the basis of the optimal solution with several rounding methods, a small variable range is given for each stacking angle of each partition, which is then rounded and optimized. Finally, the optimal solution is checked to fine-tune the final result. The whole procedure shown in figure 2.

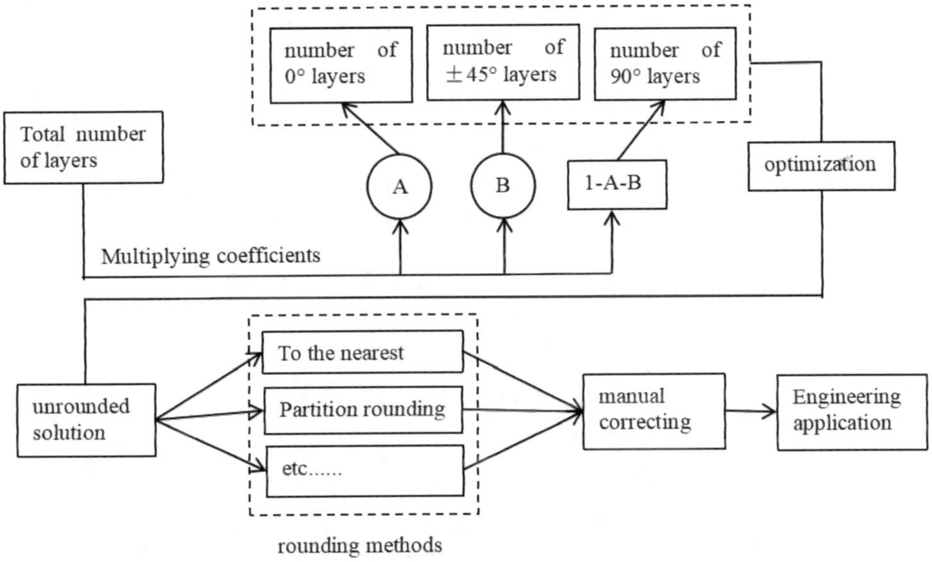

Figure 2. Quadratic rounding process.

3. Calculation Analysis

3.1. Computational model

A certain type of composite wing structure with a large aspect ratio is used as an example of optimization calculation. The model is shown in figure 3, and the main structural materials are shown in Table 2. The wing-box skin, spar, and stringer are made of composite materials, while ribs, joints, etc. are made of metal. The main thickness distribution of the upper and lower skins is shown in figure 4.

Composite wing size, stacking, angle, and other related parameters are used as design variables, and statics, stability, aeroelasticity, and other strength parameters are used as design constraints, as shown in Table 1. The lightest weight is the design goal and optimal design calculation is performed based on Nastran SOL 200.

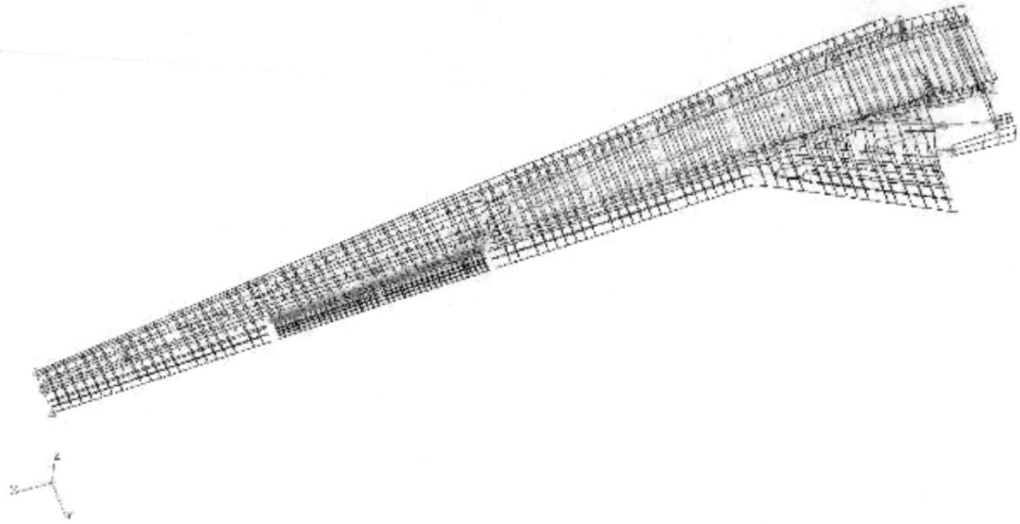

Figure 3. The finite element model of a wing structure with a large aspect ratio.

Table 1. Optimization Constraints.

Constraints	Amount
Wing tip angle (°)	≤3
Wing tip displacement (mm)	≤1500
Instability factors	≥1
(Sea level) Flutter velocity (m/s)	≥147
Permissible stress constraint of the laminate's tension-compression design με	[-4000, 5500]
Permissible stress constraint of the laminate's shearing design με	[-7600, 7600]

Table 2. Material Properties.

Materials	Properties	Amount
3234/T700	Modulus of compression at 0° (MPa)	1.1e5
	Tensile modulus at 0° (MPa)	1.2e5
	Modulus of compression at 90° (MPa)	7800
	Tensile modulus at 90° (MPa)	8700
	In-plane shear modulus (MPa)	4000
	Poisson's ratio vLT	0.32
	Density (kg/mm3)	1.7e-6
7050T7451	Elastic modulus (MPa)	70000
	Poisson's ratio v	0.33
	Density (kg/mm3)	2.8e-6
TC21	Elastic modulus (MPa)	1.09e5
	Poisson's ratio v	0.34
	Density (kg/mm3)	4.44e-6

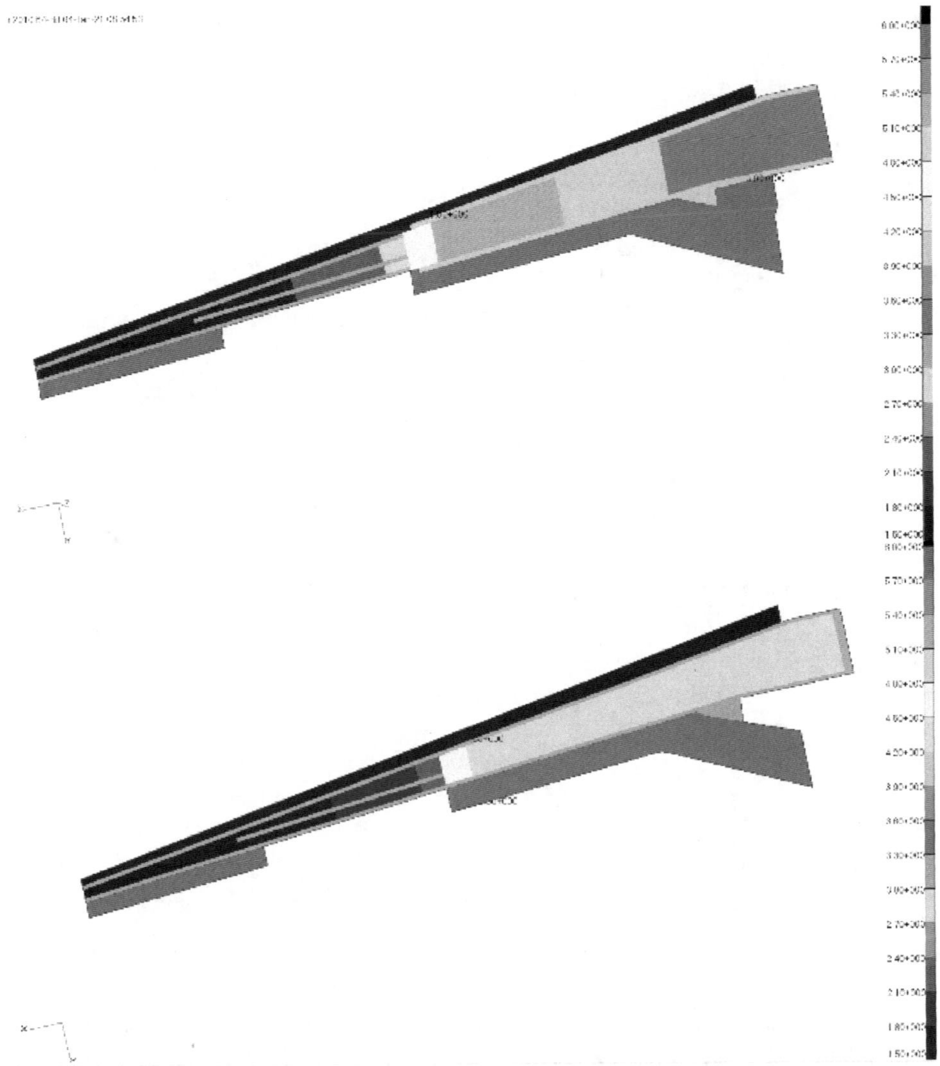

Figure 4. Thickness distribution of the upper and lower composite skins.

In static calculation and analysis, the severe working condition of the aircraft flight load is selected. The relevant parameters of the aerodynamic load and inertial load of this condition are shown in Table 3, with the 1.3 safety factor in consideration.

Table 3. Load.

Working condition	Amount	Working condition	Amount
Vertical load Fz(N)	37518	Roll acceleration (rad/s2)	0
Spanwise centre of pressure X_F(m)	5.268	Pitch acceleration (rad/s)	-67.4
Navigation centre of pressure Y_F(m)	3.82	Unilateral load (N)	23276
Unilateral mass (kg)	530.1	Unilateral bending moment (N•m)	113257
Spanwise centre of mass (m)	2.111	Overload (ny)	2.72

3.2. Calculation and comparison of the optimized wing

After the optimization analysis, this section lists the calculation results of the wing optimization model and the final values of the optimization variables and selects two optimal solutions for comparative analysis. The results of wing model changes and strength-related parameters before and after comparison are shown in Table 4, and the changes in wing weight are shown in Table 5.

Table 4. Comparison of results before and after optimization.

Subject	Calculation	Pre-optimization	Optimal solution 1	Optimal solution 2
Displacement	Vertical displacement at wing tip (mm)	1488	1511	1491
	Wing tip torsion angle (°)	2.90	2.93	2.77
Stability	First order instability factors	1.316	1.08	1.17
Modal analysis	First vertical bending (Hz)	3.27	3.24	3.22
	Second vertical bending (Hz)	9.10	9.51	9.58
	First horizontal bending (Hz)	10.36	10.34	10.32
	Inboard aileron deflection (Hz)	16.38	16.38	16.38
	Second horizontal bending (Hz)	17.93	18.17	18.16
	Outboard aileron deflection (Hz)	18.48	18.47	18.47
	Third vertical bending (Hz)	19.11	19.28	19.67
	Fourth vertical bending (Hz)	30.07	28.70	29.45
	Third horizontal bending (Hz)	32.76	33.46	33.62
	First-order torsion (Hz)	39.79	37.28	37.58
Flutter	Fluttering speed (m/s)	264	280	310

Table 5. Weight before and after optimization.

Structure	Pre-optimization	Optimal solution 1	Optimal solution 2
Total wing weight	520.32	512.00	515.48
Upper skin	32.192	26.615	29.055
Lower skin	22.796	21.910	22.621
Front spar web	6.138	7.667	7.574
Front spar flange	7.130	4.207	4.109
Rear spar web	5.269	4.866	5.093
Rear spar flange	6.184	6.123	6.417

In the stability calculation, the results of optimal solutions 1 and 2 are shown in figure 5. Compared with the original model before optimization, the wing stability is significantly improved. Since the stability of composite laminates is mainly determined by the D-matrix, the optimization of the stacking sequence makes a great contribution to the improvement of the stability of the skin panels.

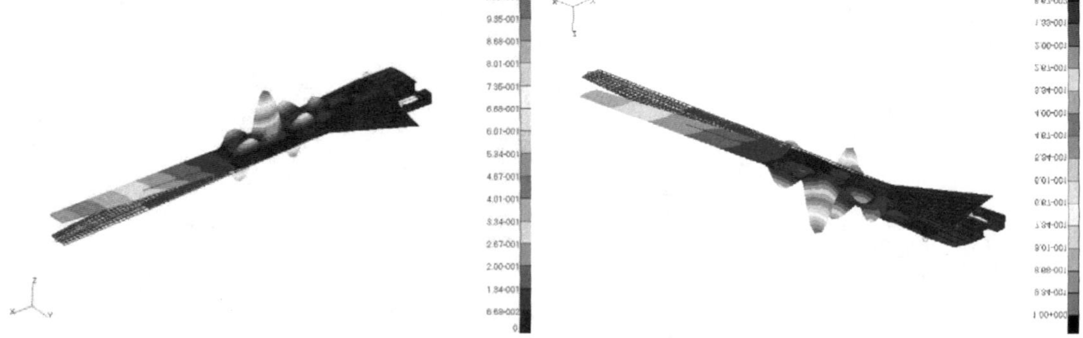

Figure 5. First-order modal instability of optimal solutions 1 and 2 (λ_1=1.08/1.17).

The flutter velocity has been improved with either optimal solution 1 and solution 2, and the V-g and V-f diagrams of the flutter analysis are shown in figure 6 and figure 7. It can be seen from the figures that the flutter velocity in optimal solution 1 is about 280m/s and that in optimized solution 2 is about 310m/s.

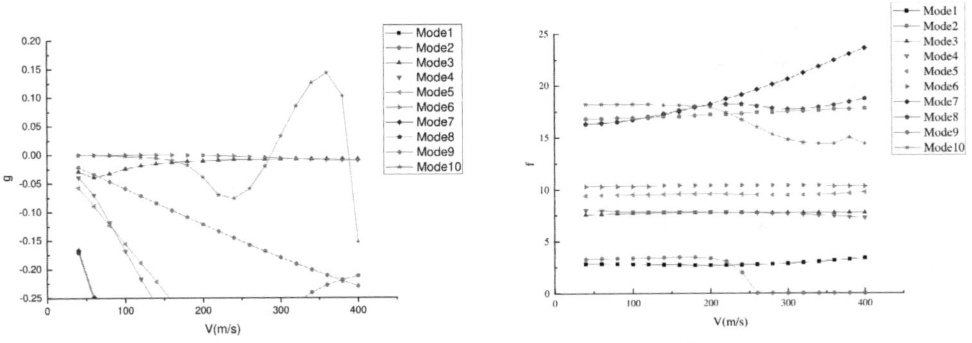

V-g diagram of optimal solution 1 V-f diagram of optimal solution 1

Figure 6. Flutter analysis of optimal solution 1.

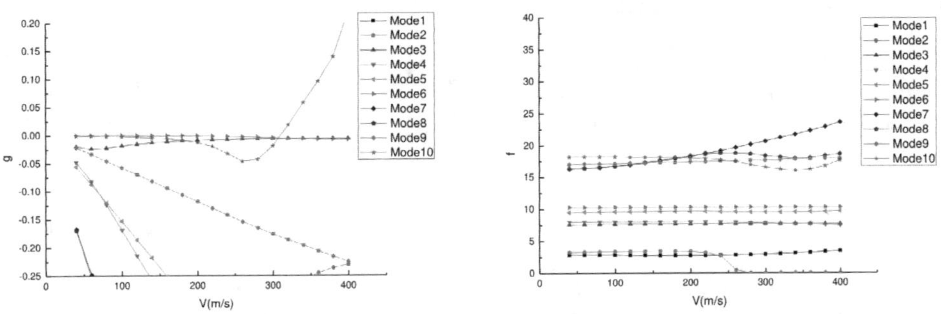

V-g diagram of optimal solution 2 V-f diagram of optimal solution 2

Figure 7. Flutter analysis of optimal solution 2.

3.3. Variable values of optimized wing

Partial optimization results of composite structure stacking variables in optimal solutions 1 and 2 are listed, which are aimed at the load-bearing area of the main wing box between the front and rear spars of the wing. The number of web partitions before and after optimization of numerical results of the web stacking is increased from 5 to 16, as shown in figure 8. figure 9 indicates the result partitions of the skin panel stacking, and Table 6 as a typical example shows the comparison of the partitions of the upper skin panel, where the original numerical results are non-integer before the quadratic rounding process.

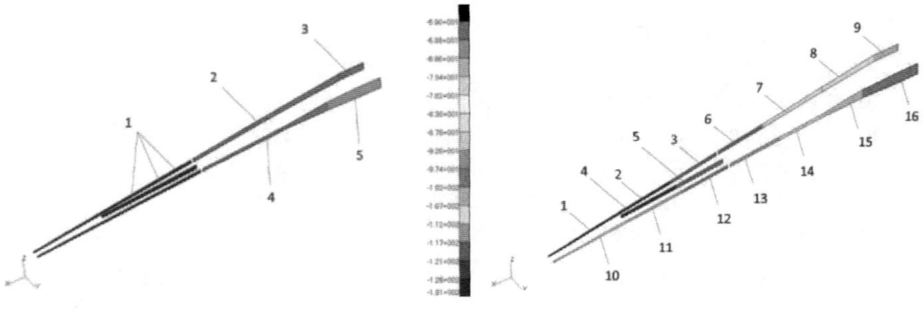

(a) (b)

Figure 8. Original stacking partition (a) and optimized partition (b) of the web.

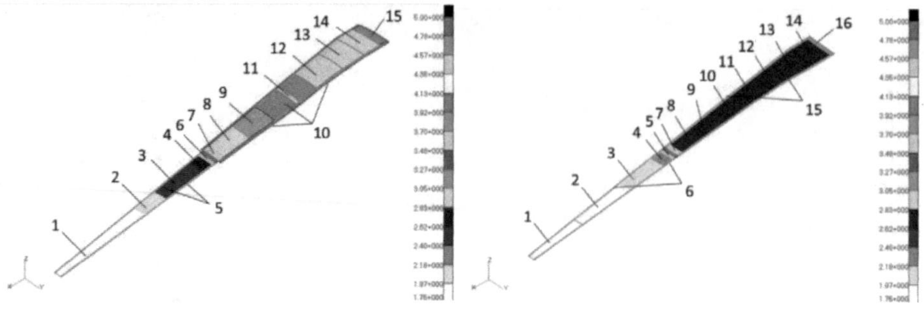

(a) (b)

Figure 9. Upper (a) and lower (b) skin panel partition.

Table 6. Comparison of the optimization results of upper skin panels.

Partition	Original stacking number	Thickness (mm)		Number of layers in the original model				Number of layers after optimization			
		Original	Optimized	0°	45°	-45°	90°	0°	45°	-45°	90°
1	14	1.75	2	4	4	4	2	6	4	4	2
2	16	2.0	2.25	6	4	4	2	8	4	4	2
3	20	2.5	2.75	10	4	4	2	8	6	6	2
4	22	2.75	3.125	8	6	6	2	9	6	6	4
5	24	3	3.5	8	6	6	4	10	6	6	6
6	36	4.5	4.875	16	8	8	4	17	8	8	6
7	36	4.5	4.875	16	8	8	4	17	8	8	6
8	24	3	1.5	8	6	6	4	6	2	2	2
9	26	3.25	2.125	10	6	6	4	7	4	4	2
10	32	4	2.5	12	8	8	4	4	4	2	4
11	36	4.5	3.75	16	8	8	4	8	8	4	8
12	28	3.5	3.375	12	6	6	4	6	6	4	6
13	28	3.5	3.125	12	6	6	4	6	6	4	6
14	28	3.5	3.125	12	6	6	4	6	6	4	6
15	32	4	3.5	12	8	8	4	6	6	6	6

4. Conclusion

In the sample calculation, the genetic algorithm is used to optimize the design of the composite wing box, and new composite combination is obtained. This calculation process attained the optimal structure of the composite wing structure with efficiency, which is summarized as follows:

1) The application of multi-objective constrained optimization method of composite wing structure is feasible and can bring a relative small weight reduction in an aircraft detail design phase comparing to the structure layout design, in where the engineering experience and hand calculation are predominantly. Through genetic algorithm, new composite wing box with more efficient and reasonable structural stiffness distribution can be obtained. In this optimized wing structure, constraints from structure strength, deformation, stability and aeroelasticity are satisfied while weight reduction is also

achieved. And in order to achieve a practical application, quadratic rounding process is necessary which performs small-scale manual rounding optimization.

2) Comparing various structure performances of the calculation examples, optimal solution 1 loses 8.32kg in weight (the total weight of the wing structure is 211.7kg, making a weight reduction of 3.93%) while its performance is equivalent to the original structure. Only the structural stability capability is lower but still within the design requirements, so it is an available optimal solution. The weight loss of optimal solution 2 is 4.84kg, but it has better structural stability. The results of optimal solution 2 can be used to replace the original model for different needs. Further peculiarities are not easy to summarized due to the high-coupled composite layers stacking stiffness, while the comparison of these two structures solution indicates that the stability factor is an important parameter affecting the weight reduction in the structure (upper skin of the wing) under compression load.

References

[1] Isogai K. Direct search method to aeroelastic tailoring of a composite wing under multiple constraints[J]. Journal of aircraft. 1989, 26(12): 1076-1080.

[2] Guo S, Cheng W, Cui D. Optimization of Composite Wing Structures for Maximum Flutter Speed[M]. 46th AIAA/ASME/ASCE/AHS/ASC Structures, Structural Dynamics & Materials Conference, 2005.

[3] Guo S, Cheng W, Cui D. Aeroelastic Tailoring of Composite Wing Structures by Laminate Layup Optimization[J]. AIAA journal. 2006, 44(12): 3146-3150.

[4] Dillinger J K S, Klimmek T, Abdalla M M, et al. Stiffness Optimization of Composite Wings with Aeroelastic Constraints[J]. Journal of aircraft. 2013, 50(4): 1159-1168.

[5] Bret K. Stanford C V J. Trim and Structural Optimization of Subsonic Transport Wings using Nonconventional Aeroelastic Tailoring[J]. 32nd AIAA Applied Aerodynamics Conference. 2014.

[6] Nagendra S, Jestin D et al. Improved genetic algorithm for the design of stiffened composite panels [J]. Computers and Structures, 1996, 58(3):543-555.

[7] Fatemi J, Trompette P. Optimization of stiffened plates using a modified genetic algorithm [R]. AIAA-98-4973,1988.

[8] Kang J H, Kim C G. Minimum-weight design of compressively loaded composite plates and stiffened panels for post-buckling strength by genetic algorithm [J]. Composite Structures, 2005, 69(2):239-246.

[9] Liu W, Butler R, Mileham A R, et al. Optimum design, experimental testing and post-buckling analysis of thick composite stiffened panel [R]. AIAA-2005-1826, 2005.

[10] Zou Congqing. The Mechanisms and Benefits of Aeroelastic Tailoring [J]. Acta Materiae Compositae Sinica. 1989(4): 1-9, 106.

[11] Zhang Yining, Yang Xu. Integrated Optimum Design of Wing Structures with Composite Skins [J]. Acta Aeronautica et Astronautica Sinica. 1997, 18(6).

[12] Wan Zhiqiang, Yang Chao, Li Zhengneng. Application of hybrid genetic algorithm in aeroelastic multidisciplinary optimization [J]. Journal of Beijing University of Aeronautics and Astronautics. 2004(12): 1142-1146.

[13] Cheng Wenyuan, Cui Degang. Optimization for composite wing based on Pareto genetic algorithm[J]. Journal of Beijing University of Aeronautics and Astronautics. 2007, 33(2).

[14] Lu Jie, Nie Xiaohua, Chang Liang. Multi-Constraint Optimization Design of Composite Wing Structure [J]. Engineering and Test, 2021, 61(2): 50-52.

[15] Liu Kelong, Yao Weixing, Yu Xiongqing. Multidisciplinary Structural Aerodynamic Design Optimization of Wings with Low Degree-of-freedom Collaborative Optimization [J]. Acta Aeronauticaet Astronautica Sinica, 2007, 28(5): 1025-1032.

[16] Zhang Tieliang, Ding Yunliang. Optimization of the Structural Layout of Composite Stiffened Panel [J]. Journal of Nanjing University of Aeronautics & Aeronautics. 2010, 42(1):8-12.

AUTHOR INDEX

Institute of Physics
Dirac House, Temple Back
Bristol BS1 6BE UK

ISSN: 1742-6588
ISBN 978-1-7138-8541-2

CURRAN ASSOCIATES INC.
proceedings.com

International Journal on Advances in Life Sciences

2016 Volume 8 Number 1&2

ISSN: 1942-2600

IARIA